ESTIMATING THE FLAMMABLE MASS OF A VAPOR CLOUD

This is one of a series of CCPS *Concept* books. A complete list of publications available from CCPS appears at the end of this book.

ESTIMATING THE FLAMMABLE MASS OF A VAPOR CLOUD

A CCPS CONCEPT BOOK

John L. Woodward

AMERICAN INSTITUTE OF CHEMICAL ENGINEERS

CCPS

CENTER FOR CHEMICAL PROCESS SAFETY
of the
American Institute of Chemical Engineers
3 Park Avenue, New York, New York 10016-5901 USA

Library of Congress Cataloging-in Publication Data
Woodward, John Lowell.
Estimating the Flammable Mass of a Vapor Cloud / John L. Woodward.
p. cm.
Includes bibliography and index.
ISBN 978-0-8169-0778-6
1. Vapors—Flammability—Mathematical models. 2. Explosions—mathematical models. 3. Fire—Mathematical models. 4. Chemical plants—Safety measures. I. Title
QD516.W66 1999 97-26844
660´.2804—dc21

CONTENTS

PREFACE

For nearly 40 years the American Institute of Chemical Engineers (AIChE) has been involved with process safety and loss control issues in the chemical, petrochemical, hydrocarbon process, and related industries and facilities. AIChE publications are information resources for the chemical engineering and other professions on the causes of process incidents and the means of preventing their occurrences and mitigating their consequences.

The Center for Chemical Process Safety, a Directorate of the AIChE, was established in 1985 to develop and disseminate information for use in promoting the safe operation of chemical process and facilities and the prevention of chemical process incidents. With the support and direction of its advisory and management boards, CCPS established a multifaceted program to address the need for process safety technology and management systems to reduce potential exposures to the public, the environment, personnel, and facilities. This program entails the development, publication, and dissemination of *Guidelines* relating to specific areas of process safety; organizing, convening, and conducting seminars, symposia, training programs, and meetings on process safety-related matters; and cooperating with other organizations and institutions, internationally and domestically, to promote process safety. Within the past several years, CCPS extended its publication program to include a "Concept Series" of books. These books are focused on more specific topics than the longer, more comprehensive books in the *Guidelines* series and are intended to complement them. With the issuance of this book, CCPS will have published over 40 books.

CCPS activities are supported by the funding and technical expertise of over 90 corporations. Several government agencies and nonprofit and academic institutions participate in CCPS endeavors.

In 1989, CCPS published the landmark *Guidelines for the Technical Management of Chemical Process Safety.* This book presents a model for process safety management built on twelve distinct, essential, and interrelated elements. The forward to that book states:

"For the first time, all the essential elements and components of a model of a technical management program in chemical process safety have been assembled in one document. We believe the *Guidelines* provide the umbrella under which all other CCPS Technical Guidelines will be promulgated."

This "Concept Series" book *Estimating the Flammable Mass of a Vapor Cloud* supports several of the twelve elements of process safety enunciated in Guidelines for the Technical Management of Chemical Process Safety, including Process Knowledge and Documentation, Process Risk Management, Incident Investigation, and Enhancement of Process Safety Knowledge. The purpose of this book is to assist designers and operators of chemical facilties to realistically estimate the fraction of a cloud or plume of accidentally-released flammable material that is capable of igniting.

ACKNOWLEDGMENTS

The American Institute of Chemical Engineers and the Center for Chemical Process Safety (CCPS) express their gratitude to all of the members of the Task Force on Flammable Mass for their generous efforts and technical contributions in the preparation of this Concept Book. CCPS also acknowledges the contributions of the Explosions and Modeling Subcommittees in recognizing a need for melding the technical approaches and knowledge in these two areas into a coherent and useful publication for chemical process professionals.

The chair of the Task Force on Flammable Mass was Robert L. Moser of ARCO Chemicals. The task force members were John Davenport, Industrial Risk Insurers (Chair, Explosion Subcommittee); Ronald J. Lantzy, Rohm and Haas Company (Chair, Modeling Subcommittee); John V. Birtwisle, Solutia Company; Donald J. Connolley, AKZO Nobel Chemicals; Randy Hawkins, Celanese Corp.; Robert E. Linney, Air Products and Chemicals Company; Robert A. Mancini, Amoco Corporation; John T. Marshall, Dow USA–Texas Operations; Larry J. Moore, Factory Mutual Engineering; and Ephraim Scheier, Occidental Chemicals Corporation. Martin E. Gluckstein was the CCPS Staff Liaison and was responsible for the overall administration and coordination of the project.

Det Norske Veritas (U.S.A) Inc. (DNV-Technica) was the contractor responsible for preparing this Concept Book. Dr. John L. Woodward was DNV-Technica's Project Leader and principal author. Ian Sheriff of the London library and Thor Foyn provided valuable assistance.

Before publication, all CCPS books are subjected to a thorough peer review process. CCPS also gratefully acknowledges the thoughtful comments and suggestions of the peer reviewers: Daniel A. Crowl, Michigan Technological University; Randy A. Freeman, Solutia, Inc.; Alan Godso, Cigna Insurance Co., David J. Hesse, Battelle Memorial Institute; Georges Melhem, A. D. Little, Inc.; Meso Molag, TNO; Kenneth Mosig, AIU Energy; John A. Noronha, Eastman Kodak Co., Hans Pasman, TNO Defense Research; Jan C. Windhorst, Nova

Chemical, Ltd.; Gene K. Lee, Air Products & Chemicals Corp.; Wilfred K. Whitcraft, DuPont Company; Kenneth W. Steinberg, Exxon Research & Engineering Co.; Craig Matthiessen, US EPA.

GLOSSARY

Accidental Chemical Release: An unintended, or sudden release of chemical(s) from manufacturing, processing, handling, or on-site storage facilities to the air, water, or land.

Adiabatic: A system condition in which no heat is exchanged between the chemical system and its environment.

Adiabatic Lapse Rate (ALR): The negative of the temperature gradient established as dry air ascends in the atmosphere. For air the ALR, λ is -0.00995°C/m (or ≈ -0.01 K/m) for heights below the maximum typical mixing height ($< \approx 2000$ m).

Advection: The transport of material by and in the wind.

Aerosol Fraction: The fraction of liquid phase, $1 - x$, which, after flashing to the atmosphere, remains suspended as an aerosol.

Along-Wind Distance, *x*: Distance in the direction the vapor cloud is traveling, i.e. the wind direction. Since the wind direction may change, the along-wind distance may change in direction and time as well.

Atmospheric Dispersion: The low momentum mixing of a vapor/gas or aerosol with air. The mixing is the result of turbulent energy exchange, which is a function of wind (mechanical eddy formation) and atmospheric temperature profile (thermal eddy formation).

Atmosperic Stability: A measure of the degree of atmospheric turbulence, particularly vertical mixing in the atmosphere. In neutral stability, the vertical temperature gradient equals the adiabatic lapse rate (ALR). Stable atmospheric conditions refer to a gradient less than the ALR (ultimately to a temperature inversion), and unstable conditions to a gradient greater than the ALR.

Autoignition Temperature: The minimum temperature of a substance required to cause self-sustained combustion in air, with no other sources of ignition.

Averaging Time: The time in atmospheric dispersion testing over which concentration data are averaged to produce a concentration-time series.

Blast Wave: The narrow pressure pulse transmitted by an explosion.

Buoyant Gas: See Positively Buoyant Gas.

Buoyant Force: Mathematically, $(\rho_{cld} - \rho_{air})gV_{cld}$ (in Newtons), the product of the difference in density of a cloud and the surrounding medium (air), the gravitational acceleration, and the volume of the cloud.

BLEVE (Boiling-Liquid-Expanding-Vapor Explosion): A type of rapid phase transition in which a liquid contained above its atmospheric boiling point is rapidly depressurized, causing a nearly instantaneous transition from liquid to vapor with a corresponding energy release. A BLEVE of flammable material is often accompanied by a large aerosol fireball, since an external fire impinging on the vapor space of a pressure vessel is a common cause. However, it is not necessary for the liquid to be flammable to have a BLEVE occur.

Computational Fluid Dynamics Models: Dispersion models that treat the partial differential Navier–Stokes equations describing fluid flow by numerically solving them using a grid of nodes or finite elements.

Concentration: The relative amount of a substance when combined or mixed with other substances. Concentration can be expressed as mole fraction, mass fraction, or component mole or mass density. Examples: y_i = 2 ppm mole fraction; w_i = 0.50 mass fraction; c_i = 0.03 kmole/m^3.

Condensed-Phase Explosion: An explosion that occurs when the material is present in the form of a liquid or solid.

Confined Explosion: An explosion of a reacting mass inside a closed system (e.g., a vessel or building).

Confinement: Obstacles such as walls and ceilings of a building, vessel, pipe, etc. that serve to limit the expansion of a dispersing or exploding vapor cloud.

Congestion: Small obstacles in the path of a flame front or dispersing vapor cloud that serve to deflect the flame front or cloud and to generate turbulent mixing.

Continuous Release: Emissions that are long in duration compared with the travel time (time for cloud to reach a location of interest) or averaging time.

Decomposition: Breakdown of material or substance (by heat, chemical reaction, electrolysis, decay, or other processes) into parts or elements or simpler compounds.

Deflagration: The chemical reaction of a substance in which the reaction front advances into the unreacted substance at less than sonic velocity. Where a blast wave is produced that has the potential to cause damage, the term *explosive deflagration* may be used.

Dense Gas: A gas with density exceeding that of air at ambient temperature. Sometimes the expression is used loosely to apply to aerosols such as flashing liquid ammonia. See also heavy gas and negatively buoyant vapors.

Detonation: A release of energy caused by the extremely rapid chemical reaction of a substance in which the reaction front advances into the unreacted substance at greater than sonic velocity.

Duration: The length of time for which an event occurs.

Energy Scaling Factor: The cube root of the TNT equivalent mass, $W^{1/3}$. [See Eq. (4.76).]

Entrainment: The mixing of air into a vapor cloud.

Expansion Ratio, r_v: The ratio of the volume after combustion to that prior to combustion.

Explosion: A release of energy that causes a blast wave.

Explosion Efficiency, ε: The ratio of the mechanical energy released in an explosion to the heat of combustion times the flammable mass in a vapor cloud (net efficiency). Alternately, the ratio of the mechanical energy released in an explosion to the heat of combustion times the total mass of fuel in a vapor cloud (gross efficiency).

Explosion Vent: An intentionally weakly supported panel in the wall of an enclosure designed to give way in the event of an explosion in order to reduce the explosion overpressures and thereby protect the rest of the structure.

Flame Expansion Dimensions: The dimensions in which a burning or exploding gas is free to expand, 1D, 2D, or 3D.

Flammable Limits: The minimum and maximum concentrations of a combustible material (gas, vapor, aerosol, or dust) in air that will propagate a flame. See LFL and UFL.

Flammable Mass: The mass of fuel in a vapor cloud that is in the flammable range, m_f.

Flammable Range: The range of concentrations of a fuel in air that will ignite with a strong ignition source (at a certain probability or percent of trials).

Flash Point: The minimum temperature at which a liquid gives off vapor in sufficient concentration to form an ignitable mixture with air near the surface of the liquid as determined by a specific test method, i.e., open-cup or closed-cup.

Full Volume Deflagration (FVD): A deflagration explosion developed from a flammable cloud of gas, aerosol, or dust that occupies the entire volume in a confined space.

Gaussian Model: A dispersion model based on the concept that atmospheric diffusion is a random mixing process driven by turbulence in the atmosphere. The concentration at any point downwind of a point-source continuous release is approximated by a concentration profile of the form $\exp(-0.5y^2/\sigma_y^2)$ in both the horizontal and the vertical dimensions. There is also a line-source Gaussian solution and a point-source instantaneous release Gaussian solution.

Gravity Slumping: The decrease in cloud height of a flowing dense gas due to the effects of gravity (negative buoyancy).

Heat of Combustion, H_{ci}: The heat of reaction obtained by burning a unit mass of a substance at the stoichiometric concentration in air to produce fully oxygenated products. If water is a product, then the "gross" heat of combustion includes the heat of condensation of water, so that liquid water is a product. Correspondingly, the "net" heat of combustion does not include the heat of condensation of water, and vapor water is a product. Unless otherwise stated, we refer here to "net" heat of combustion.

Heat of Reaction: The net difference in heat of formation of all reactants and of all products in an adiabatic system. The reaction is exothermic if heat is released (heat of reaction is negative), and endothermic if heat is absorbed by the reaction.

Heavy Gas: A gas with density exceeding that of air at ambient temperature. See also dense gas and negatively buoyant vapors.

Ignition Source Density: The number of ignition sources per unit area (of plant).

Impulse, *I*: The area under the overpressure-time curve for explosions. The area can be calculated for the positive pressure phase or the negative phase of the blast wave.

Insolation: Solar radiation heat loading (Watts/m^2).

Instantaneous Release: Emissions that are short in duration compared with the travel time (time for cloud to reach a location of interest) or compared to the averaging time.

Integral Model: A dispersion model which averages or "integrates" the concentration in a given dimension or time so that concentrations can be described by solving an ordinary differential equation instead of a partial differential equation.

Intermittency: The fraction of the time that a measured value is zero in a time-series of measurements.

Jet Discharge: A release of vapor or aerosol at sufficient pressure that the momentum of the release provides the dominating mechanism for air entrainment and for the centerline trajectory of the release.

Jet Fire: A type of fire occurring with a pressurized release of gas and/or liquid.

Lower Flammable Limit (LFL): The lowest concentration of combustible material that will propagate a flame from an ignition source through a mixture of flammable gas, aerosol, or combustible dust in air. Also known as the Lower Explosive Limit (LEL).

Mach Number, M_w: The ratio of the speed of a flame or blast wave to the speed of sound at ambient temperature and pressure.

Minimum Flammable Mass: The lowest mass of fuel in the flammable range that will cause a defined level of damage.

Minimum Ignition Energy: The lowest energy of an electrical spark discharge that will ignite a flammable mixture of fuel and air under defined test conditions.

Momentum Flux: Mathematically, ρu^2 (in Newtons/m^2), the product of the cloud density and speed squared.

Momentum Force: Mathematically, $\rho u^2 A$ (in Newtons), the product of the cloud density, speed squared, and cross-sectional area.

Negatively Buoyant Vapors: Vapors or aerosol with a specific gravity greater than ambient air so the vapors tend to sink to low spots and to spread along the ground by "gravity spreading."

NFPA: Acronym for the National Fire Protection Association (U.S.).

Nose of Vapor Cloud: The front surface of vapor cloud, that is, the surface farthest along in the along-wind direction. All surfaces of a vapor cloud are usually defined by a certain concentration such as the LFL.

Overpressure, ΔP: The increase of pressure above ambient in an explosion wave.

Partial Volume Deflagration (PVD): A deflagration explosion developed from a flammable cloud of gas, aerosol, or dust that occupies only part of the volume in a confined space.

Passive Dispersion: Dispersion caused by the normal turbulence in the atmosphere. See Gaussian model.

PES: See Potential Explosion Sites.

Physical Explosion: The catastrophic rupture of a pressurized gas/vapor-filled vessel by means other than reaction, or the sudden phase-change from liquid to vapor of a superheated liquid.

Pool Fire: The combustion of material evaporating from a layer of liquid at the base of the fire.

Positively Buoyant Gas: A gas with density less than that of air at ambient temperature.

Potential Explosion Site (PES): A volume within a plant with sufficient congestion and/or confinement that a flammable vapor cloud ignited there could likely develop into an explosion.

Process Safety Management: A program or activity involving the application of management principles and analytical techniques to ensure the safety of chemical process facilities. Sometimes called process hazard management.

Puff Release: See instantaneous release.

Rain Out: When a superheated liquid is released to the atmosphere, a fraction of it will flash into vapor. Another fraction may remain suspended as an aerosol. The remaining liquid, as well as portions of the aerosol, η_R, may fall or "rain out" onto the ground.

Reactants: Chemicals that are converted into products during the reaction process.

Reaction: The process in which some chemicals (reactants) are converted into other chemicals (products).

Reaction Kinetics: The factors that determine the rate of reaction.

Reactivity: The relative tendency of a substance to undergo chemical reaction (low, medium, or high).

Risk: A measure of potential economic loss, human injury, or environmental damage (cost) in terms of the probability of the loss, injury, or damage over a period of time, normally a year. Plotting the cost vs. probability gives one representation. The integral under this curve gives another, called the Societal Risk.

Runaway Reaction: A thermally unstable reaction system which shows an accelerating increase of temperature and reaction rate.

Sachs' Scaling: Dimensionless terms for TNT equivalence explosion modeling. See Table 4.9.

Screening Tool: A simplified dispersion model with limited capabilities, suitable for screening-level studies.

Shallow-Layer Model with Terrain Effects: A dispersion model capable of predicting flows in variable terrain by using a simplification in the model formulation that makes the equations applicable only to a shallow layer of fluid.

Side-on Overpressure: The level of overpressure in the pressure wave from an explosion measured perpendicular to the direction of propagation of that pressure wave.

Standard Deviation (cloud concentration), σ: The square root of the sum of squared values of concentration minus centerline concentration. The cloud half-width can be defined as the distance corresponding to a concentration of two or three standard deviations from the centerline.

Stand-Off Distance, *R*: Distance from the center of an explosion.

Stratification: As used here, any concentration profile that develops in an enclosure. The flammable mass will be between the LFL and UFL concentrations along such a concentration profile.

Superheated Liquids: Liquids at a temperature above their normal boiling point.

Top-Hat Dispersion Models: Dispersion models with no concentration gradients within the vapor cloud. The concentration is modeled as a well-mixed system within the vapor cloud and zero outside the cloud.

TNT Equivalence Models: Models of explosion characteristics relating vapor cloud explosions to the explosion of an equivalent mass of TNT (trinitrotoluene).

Unconfined Vapor Cloud Explosion: An explosion in a vapor cloud that is outdoors and not confined by walls and a ceiling on all sides. This term is not recommended and is not used in this book, since explosions require some degree of confinement or congestion or a strong directed source.

Upper Flammable Limit (UFL): The highest concentration of combustible material that will propagate a flame from an ignition source through a mixture of flammable gas, aerosol, or combustible dust in air. Also known as the Upper Explosive Limit (UEL).

Vapor Pressure, P_s: The pressure exerted by a saturated vapor in equilibrium with its own liquid.

Volume Blockage Ratio (VBR): The ratio of the volume occupied by congestion elements such as pipes, beams, plates, etc. to the volume of the portion of the plant under consideration.

Volume-Source Explosion Models: Models that predict explosion characteristics based on the volumetric portion of the flammable cloud involved in the explosion (that portion of the flammable cloud that is influenced by congested and/or confined volumes in a plant).

NOMENCLATURE

a_0	= speed of sound at ambient conditions, m/s
a,b	= regression constants in Eq. (4.80) or (5.10)
a_{cd}	= correlation constant for gas or dust explosions
A	= cross-sectional area of puncture, m^2
A	= area in horizontal plane of plant over which obstacle effect on surface roughness length is evaluated, m^2 in Eq. (4.48)
A	= plan view area of flammable cloud, m^2 in Eqs. (4.81)–(4.84) and (5.11)
A	= kinetics premultiplier in Eq. (4.51)
A_f	= area swept by a flammable cloud dispersed to its full extent, m^2 in Eqs. (4.81)–(4.84)
A_j	= jth surface area for contaminant deposition in Appendix G, m^2
AFT	= adiabatic flame temperature, K
b_{0y}	= half width of source, m
b_{0z}	= half height of source, m
B_c	= buoyancy flux defined in Appendix H
c	= mean concentration, kg/m^3
ch_i	= number of chains of carbon atoms on a fuel molecule
c_i	= indoor concentration, kg/m^3
c_o	= source concentration or outdoor concentration, kg/m^3
C_D	= discharge coefficient (—)
C_p	= heat capacity at constant pressure, J/(kg K)
C_v	= heat capacity at constant volume, J/(kg K)
d,e	= constant parameters
d_p	= drop diameter, m
d_N	= diameter of nozzle or orifice, m
d_{32}	= Sauter mean diameter, m
D	= average obstacle size, m

D = diameter of vessel or pipe, m
D_h = puncture diameter, m
D_p = diameter of pool, m
E = energy in explosion, J
E = activation energy for reaction, J/kgmole
f = Fanning friction factor, (—)
f_m = mass fraction of released fuel in flammable region, (—)
f_r = fraction of the distance between the release height and the ceiling
f_{SP} = mass fraction in spray (not raining out)
f_{z1}, f_{z2} = functions used to express $\sigma_z(x)$
F = swelling factor of fuel, m^3/m^3
F_I = pipe inclination factor defined by Eq. 4.10, (—)
F_0 = mass fraction of contaminant removed by air filter, (—)
F_1 = mass fraction of contaminant in recirculated air removed by filter, (—)
F_x = along-wind concentration profile for instantaneous puff, (—)
F_y = cross-wind concentration profile, (—)
F_z = vertical concentration profile, (—)
g = gravitational constant, m/s^2
g'_0 = buoyancy-adjusted gravitational constant, m/s^2, see Eq. (H.4)
G = mass flux of discharge, $kg/(m^2s)$
G^* = dimensionless mass flux of discharge, (—)
G_{evap} = mass flux for evaporation from a pool, $kg/(m^2s)$
G_L = liquid discharge mass flux, calculated with orifice equation, Eqs. (4.3), (4.22)
G_p = mass flux discharge from pipe, $kg/(m^2s)$
h = release height, m
h = average depth of pool, m
h = liquid head above puncture point, m
h_{mi} = effective height of vapor cloud section i, m
h_i = height of inversion layer, m
h_{min} = minimum average depth of pool at maximum pool diameter, m
h^* = average obstacle height, m in Eq. (4.48)
h_1 = height of well-mixed layer from indoor release, see Appendix H, m
h_2 = height of linear gradient layer from indoor release, Appendix H, m
H = specific enthalpy, J/kg
H = height of building or room, m
H_{fg} = heat of vaporization, J/kg
H_c, H_{ci} = specific heat of combustion of component i, J/kg

H_{comb} = total heat of combustion in flammable cloud, J
H_{cv} = heat of combustion per unit volume, J/m^3
H_v = heat flux from the earth, W/m^2
i_s = scaled impulse from explosion, see Table 4.7
I = impulse from explosion, Pa s
j,k,l,m,n = power law coefficients
k = Von Karman constant, value = 0.40 (—) in Eq. (4.46)
k = polytropic coefficient, slightly less than γ (—)
k_T = thermal conductivity of substrate (ground or water), W/(m K)
k = turbulent energy generation rate in k-ε models
k_c, k_p = parameter in Elliptic integral of second kind for continuous and instantaneous releases, respectively (—)
K_e = equivalent number of velocity heads to account for fittings, bends, expansions, etc.
K_g = reactivity of gas expressed as maximum rate of pressure increase times the cube root of the test volume (MPa m/s).
K_{st} = reactivity of dust expressed as maximum rate of pressure increase times the cube root of the test volume (MPa m/s).
L = Monin-Obukov turbulence scale, *m*, in Eq. (4.46) and Eq. (A.3)
L = length of pipe or length of room or characteristic length of surface, m
L_{ave} = average carbon chain length, number of carbon atoms
L_e = estimated eddy size discussed in Section 4.2.6, m
L_f = maximum flame path length in congested area, m
LFL = lower flammable limit, mole fraction of fuel
LNG = liquefied natural gas
LPG = liquefied petroleum gas, mostly propane and propylene
L_r = length of room, m
m_f = mass of flammable material in flammable range in vapor cloud, kg
m_t = total mass of flammable material discharged, kg
m_f = partial order of reaction with fuel component in Eq. (4.51)
m_0 = partial order of reaction with oxygen in Eq. (4.51)
M_c = momentum flux defined in Appendix H
M_i = molecular weight of component *i*, kg/kgmole
M_T = molecular weight of multicomponent fuel–air mixture, kg/kgmole
M = number of methyl groups in a fuel molecule
M_w = Mach number for flame speed, ratio of flame speed to speed of sound in unburned mixture at ambient temperature.
n_i = stoichiometric index in reaction
n_c = exponent in Eq. (4.40)

n = exponent in definition of toxic dosage; see Section 4.2.4
N = number of velocity heads lost in pipe by line friction, bends, fittings, see Eq. (4.9)
N_i = number of carbon atoms in a chain
Nu = Nusselt number, hd_p/k_T (—) where h is the heat transfer coefficient, W/(m^{2}K)
N_{eq} = nonequilibrium term defined by Eq. (4.25)
p = exponent in power law wind profile, Eq. (4.47)
$p(c,x)$ = perimeter length of concentration isopleth, m
P = probabililty of exceeding the limiting concentration when the mean concentration is $\bar{c}$, see Eq. (4.72)
P = pressure, Pa
PES = potential explosion site
P_a = ambient pressure, Pa
P_0 = initial stagnation pressure in vessel or pipe, Pa
P_s = saturation vapor pressure, Pa
q_0 = volumetric rate of forced-convection fresh air to a building, m^3/s
q_1 = volumetric rate of air recirculation in a building, m^3/s
q_2 = volumetric rate of air infiltrating a building, m^3/s
q_3 = volumetric rate of air leaving a building through leakage, m^3/s
q_4 = volumetric rate of forced-convection exhaust air from a building, m^3/s
Q = heat flux, W/m^2
Q = probability of nonignition in Eqs. (4.82)–(4.83)
Q_f = probability of nonignition of vapor cloud over the area swept by the cloud diluting to its maximum extent
r = pool radius, m
r_f = rate of fuel oxidation, kg fuel/s
r_m = mole ratio of products of combustion to reactants, see Eq. (4.59)
r_N = nozzle or orifice radius, m
R = radius or characteristic height of vapor cloud, m
R = gas constant, 8314.3 J/(kgmole K)
R = sink for contaminant (by deposition or decomposition), kg/s in Eq. (G.1)
Ri = Richardson number, see Eq. (A.1)
Ri_c = Release Richardson number, see Eq. (5.3)
Ri_v = Inlet Richardson number defined by Eq. (H.10)
s = silhouette area of all obstacles in the area, Eq. (4.48)
s = indoor source rate for contaminant, kg/s in Eq. (G.1)
S = entropy, J/(kg K)
S_{fg} = entropy difference between vapor and liquid, $S_G - S_L$, J/(kg K)

S_u = laminar flame speed, m/s
S_t = turbulent flame speed, m/s
t = time, s
t_a = arrival time of leading edge of explosion pulse, s
t_d = duration of positive overpressure pulse of explosion, s
t_i = turbulence time scale, s
T = temperature, K
T_{as} = adiabatic saturation temperature, K, see Figure 4.12
T_b = normal boiling point, K
ΔT_{st} = temperature rise in a flame at stoichiometric concentration, K
u = velocity in jet or plume, m/s
u_d = mass average drop settling velocity, m/s
u_w = wind speed, m/s
u_* = friction velocity, a parameter in vertical wind speed profiles, m/s
UFL = upper flammable limit, mole fraction of fuel
v = specific volume of single-phase or possibly two-phase mixture, m^3/kg
v_v = specific volume of vapor or gas, m^3/kg
V = cloud volume or room volume, m^3
V = volumetric rainout rate to pool, m^3/s in Eq. (4.38), (4.39)
V = velocity in y direction, m/s
VBR = volume blockage ratio (—)
w = discharge rate, kg/s
w_c = effective width of vapor cloud, m
w_p = instantaneous mass discharge, kg
w_i = mass fraction of component i, kg/kg
w_{vi} = vapor mass fraction of component i, kg/kg
W = equivalent mass of TNT from explosion, kg of TNT
W = width of building or room, m
$W(x)$ = characteristic half width of vapor cloud, m
We = Weber number defined by Eq. 4.28
We_c = critical Weber number for droplet breakup, value = 12 to 22.
x = along-wind distance, m
x_0 = along-wind distance to center of cloud, m
x = mass fraction of vapor or gas in two-phase mixture, kg/kg
X = surface correction factor in Eq. (4.41)
X_r = fraction of the volume of an enclosed room occupied by flammable cloud
y = across wind direction or radial distance of jet impacting flat surface, m

y_i = mole fraction of component i in the vapor phase, mole i/mole total
z = vertical distance, m
z_0 = surface roughness length, m
z_r = release height for pipe break inside a room, m
z_2 = vertical distance from impacting surface for jet impaction, m
Z = scaled radial stand-off distance from center of explosion.

Subscripts

air = air
a, amb = ambient
basis = time basis for averaging time, s
b = burned or combustion products
c = choked or heat of combustion
chem = heat generated by chemical reaction (oxidation)
cl = centerline
cld = vapor cloud
cls = flammable mass in a slice of the vapor cloud of width Δx
cond = heat flux by conduction
conv = heat flux by convection
cs = combustible component at saturation
cv = volumetric heat of combustion
dis = discharge
dur = duration of passing vapor cloud, s
erm = equilibrium rate model
evap = heat lost by evaporation
f = fuel or flammable portion
fg = difference between vapor and liquid (i.e., heat of vaporization or density difference)
g = gas or vapor
gross = explosion efficiency based on total mass released
H = evaluated at constant enthalpy
i = index to components in reaction or to potential explosion site
ideal = ideal conditions, discharge rate with $C_D = 1$
jet = dimensions of jet
leak = enthalpy added to pool from rained-out liquid
loss = heat loss to surroundings
L = liquid
lim = limit value for concentration (LFL or UFL or fraction thereof)

LFL = distance or time to lower flammable limit
m = flammable mass or mole ratio of combustion products to reactants
max = maximum
min = minimum
mi = flammable cloud section i
net = net heat flux to/from a pool or net heat of combustion (to vapor products)
net = explosion efficiency based on flammable mass in a vapor cloud
0 = initial value at origin, stagnation point
orifice = discharge rate from an orifice
out = conditions at discharge point
O = oxygen mole fraction
p = logarithmic standard deviation of concentration fluctuations; impaction plate (surface)
p = pipe
pool = pool
P = evaluated at constant pressure
r = index to reactant
rad = radiation
s = saturated conditions (vapor pressure)
sh = superheat defined by Eq. (4.31)
sol = solar heat
st = stoichiometric concentration
st = branch crossing point in omega method discharge rates, see Eq. (4.19)
S = evaluated at constant entropy
St = stoichiometric concentrations
UFL = distance or time to upper flammable limit
v = vapor or volume expansion ratio
v, vx = virtual source distance
vz = virtual source height
wall = profile evaluated at the wall
x = along-wind value for (semi) continuous release
xp = along-wind value for instantaneous release
y = across-wind value for (semi) continuous release
yp = across-wind value for instantaneous release
z = vertical value for (semi) continuous release
zp = vertical value for instantaneous release
1 = discharge plane
2 = distance from impingement surface to point on impacted plume

Superscripts

m_f = partial reaction rate order for fuel
m_0 = partial reaction rate order for oxygen

Greek

α = thermal diffusivity of substrate, fuel droplet or particle, $k_T/(\rho C_p)$, m^2/s
α = angle from the horizontal for jet impacting flat surface, radians
α = half width of ellipse, m
α_0 = volumetric void fraction, m^3 gas or vapor/m^3 of two-phase mixture
α,β = coefficients discussed in Section 4.2.6
α = ratio q_1/q_0 of recirculated air to fresh makeup air rates in Appendix G
β = coefficients used to calculate stoichiometric mole fraction in Eq. (4.55), (4.56)
β = half height of ellipse, m
β = ratio of puncture diameter to vessel or pipe diameter, D_h/D
β = ratio q_2/q_0 of infiltrated air to fresh makeup air rate in Appendix G
γ = isentropic constant, ratio of heat capacities, C_p/C_v
γ = ratio of surface area for deposition to fresh makeup air rate in Appendix G
γ_i = argument in approximation to Elliptical integrals of the second kind
Γ = correlating parameter for gas or dust explosions
δ = sum of ratios defined in Appendix G
δ = additional mixing layer caused by momentum, defined in Appendix H
Δ = difference
ε = emissivity of flame or fuel droplet or particle, (—)
ε = explosion efficiency or yield factor, J in explosion/J in combustion energy
ε = specific volume ratio, v/v_0
ε = sum of ratios defined in Appendix G
ε = turbulent energy decay rate in k-ε models
η = pressure ratio to initial stagnation pressure in tank or pipe, P/P_0. In particular it is usually evaluated at $P = P_{out}$
η_R = mass fraction rainout, kg/kg
θ = angle from the horizontal between tank and pipe break point, radians
θ = potential temperature defined in Appendix A
θ_i = normalized indoor concentration in Appendix G

κ = thermal conductivity of fuel, W/(m,K)
λ = constant in Eq. (4.40), adiabatic lapse rate –0.0095°C/m.
λ = mole ratio of oxygen defined by Eq. (4.63)
λ = adiabatic lapse rate, ≈ -0.01 K/m used in Appendix ·
μ = constant in probability of nonignition, Eq. (4.83)
ρ = density of possibly two-phase mixture, kg/m3
ρ_0 = initial density in vessel or pipe, kg/m^3
$\tilde{\rho}$ = molar density of vapors, kgmole/ m^3
σ = Stefan-Boltzmann constant (5.67×10^{-8} Wm^{-2}K^{-4})
σ = surface tension at jet expansion conditions, Pa.s
σ = ratio of contaminant source to rate of forced convection into room, Appendix G
σ_p = standard deviation of concentration fluctuations about a mean, see Eq. (4.72)
σ_y = standard deviation of cross-wind Gaussian concentration profile, m
σ_v = turbulent velocity, m/s
σ_x = standard deviation in wind direction (for puff), m
σ_y = standard deviation in cross-wind direction, m
σ_z = standard deviation in vertical direction, m
τ = time constant for indoor concentration buildup, see Appendix G
τ_a = scaled arrival time for leading edge of explosion pulse, s/kg$^{1/3}$ or dimensionless
τ_d = scaled duration of positive overpressure pulse of explosion, s/kg$^{1/3}$ or dimensionless
τ_1, τ_2 = averaging times, s
φ = equivalence ratio
φ = azimuthal angle from jet directed toward flat surface, radians
Ψ = Function of Monin–Obukov length in Eq. (4.46) and Appendix B
ω = constant in Eq. (4.16) relating specific volume ratio to inverse pressure ratio
Ω = overturning number $=M_c/B_c$, defined in Appendix H

1

INTRODUCTION

He who anticipates the future can safeguard the present.
—John Bond

1.1. Why Calculate Flammable Mass?

Technology-based society requires handling and controlling hazardous materials. For example, gasoline, diesel fuel, and jet fuel are widely distributed and move our transportation system. Natural gas, propane, or heating oil flows into nearly all of our homes.

Occasionally the hazard inherent in employing these volatile servants is manifest in spectacular and costly fires and explosions. Such unexpected events remind us of the need to continually improve our understanding of hazardous materials, and with this understanding improve our technology for safely handling and controlling them. For example, recent research has taught us that vapor cloud explosions occur when flame speeds accelerate in congested areas or near a source of mixing such as a jet release. Conversely, in an open unconfined area flame speeds remain low, resulting in a flash fire. Knowing this, we can better estimate explosion energies and refine design parameters for locating or hardening plant buildings. Refining designs is aided by the ability to quantify the phenomena to be controlled—in particular to mathematically model them. Calculating the flammable mass in a vapor cloud is an important step in such modeling.

1.2. How Are Flammable Mass Estimates Used?

Plant designs increasingly address issues of what can go wrong. "What if" scenarios are postulated and consequences estimated to facilitate cost/benefit decisions in the choice of protective designs, instrumentation, process control, inventories stored on-site, separation distances between units, use of deluge systems, redundancy and routing of fire water systems, etc. As described, for example, in the CCPS book *Guidelines for Evaluating Process Plant Buildings for*

External Explosions and Fires, and in the API 752 document "Management of Hazards Associated with Location of Process Plant Buildings," the design and/or location of new buildings can be made rationally considering the potential for the event to occur.

Flammable mass calculations are also pivotal to the calculations of quantitative risk analysis (QRA), risk management plans (RMP) and emergency response planning. QRA typically uses many scenario cases, combining consequence estimation with event frequencies to calculate risk contours, societal risk, and financial risk for a plant or transportation system.

It is not realistic to use the total amount of flammable materials discharged for such estimates. Flammable mass estimation should be based on sound and realistic physical principles and historical experience. Analysis of possible accident scenarios or investigation of accidents usually begins with calculating estimates of the mass of flammable material released. The next step is to calculate the air entrainment rates and concentration gradients that develop. With this information we can find the fraction of the released mass that may be in the flammable region. Often this is a small fraction of the total released. Only the mass within or near the flammable limits participates in an explosion or flash fire. Only the flammable mass that is in or near a source of turbulence will participate in an explosion. The balance could feed a sustained jet or pool fire or dissipate harmlessly without igniting.

1.3. Other CCPS Publications

This "Concept Series" book supports a number of CCPS publications, as sampled below.

A systematic approach to process safety management is set out in the CCPS publications

- *Guidelines for Technical Management of Chemical Process Safety*
- *Plant Guidelines for Technical Management of Chemical Process Safety*
- *Guidelines for Implementing Process Safety Management*

Flammable mass estimation is important in four of the twelve elements of process safety defined in these guidelines: process knowledge and documentation, process risk management, incident investigation, and enhancement of process safety knowledge.

A series of books support the process safety management process by providing technical details. Of particular relevance are the following:

- *Guidelines for Evaluating the Characteristics of Vapor Cloud Explosions, Flash Fires, and BLEVEs*—Summarizes and evaluates experimental and theoretical research and methods to estimate consequences from fire and explosion events. The methods described require as crucial inputs the estimation of flammable mass in a vapor cloud.

- *Guidelines for Evaluating Process Plant Buildings for External Explosions and Fires*—Addresses fire and explosion risks to process plant buildings. Offers guidance on assessing, screening, and managing risks associated with building design and siting. Scenarios require flammable mass estimation.
- *Guidelines for Hazard Evaluation Procedures, 2nd Edition with Worked Examples*—Hazard evaluation techniques identify the situations and scenarios that can potentially result in a flammable vapor cloud release.
- *Guidelines for Chemical Process Quantitative Risk Analysis*—Shows how to calculate quantitative risk estimates for operation of process plants and transportation systems. QRA combines event frequencies and event consequences. Consequence analysis of fires and explosions requires the estimation of flammable mass.
- *Guidelines for Chemical Transportation Risk Analysis*—Extends the concepts of quantitative risk analysis to treat transportation systems (rail, road, barge, etc.). Flammable mass estimation needed in analyzing the risk of transporting flammable materials.
- *Guidelines for Use of Vapor Cloud Dispersion Models,* 2nd Edition— Provides an introduction to dispersion modeling techniques and a review of available dispersion models. Dispersion models are useful for predicting flammable mass in vapor clouds.
- *Guidelines for Investigating Chemical Process Incidents*—Describes techniques to find the cause of a process plant or transportation accident. Flammable mass estimates are usually desired with accidents involving a flammable vapor cloud.
- *Guidelines for Postrelease Mitigation Technology*—Presents engineering methods used to minimize the consequences of releases of hazardous materials such as water sprays, foams, containment, detection and response, and plugging techniques. Some of these methods need flammable mass estimates as inputs.
- *Guidelines for Vapor Release Mitigation*—Provides practical methods for minimizing a chemical release and for mitigating such an emergency. Quantification of the fire and explosion hazard requires estimating flammable mass.
- *Guidelines for Technical Planning for On-Site Emergencies*—Outlines four basic components of emergency planning: prevention, preparedness, response, and recovery, including recognizing credible incidents and planning practical strategies to deal with these incidents. Estimating flammable mass can be used in incident planning.

- *Guidelines for Engineering Design for Process Safety*—Discusses how to select designs that prevent or mitigate the release of flammable materials. To quantify the benefit of proposed process changes it is important to be able to estimate the flammable mass in releases.

2

OVERVIEW

Experience is a good teacher but the fees are heavy; ... first, an accident has to occur before we can learn from it. The second drawback of experience is that it is not readily transferable.
—Kletz, 1990

2.1. Context

Flammable mass calculations are needed to quantify the source strength estimates for fire and explosion calculations as part of a plant or transportation risk analysis. This chapter briefly sets flammable mass calculations in the context of the first few steps in risk analysis, usually termed consequence analysis.

The typical process for evaluating process and transportation risks is illustrated in **Figure 2.1**.

In Step 1 the analyst *identifies* and inventories the hazardous materials in the process. A hazard is a *potential* for harm. For example, a mountain or a cliff is a hazard since a hiker could fall over the edge or an airplane could crash into it. Risk is more than the presence of a hazard; it is the product of the consequences or cost of an undesirable event and the frequency or probability of that event.

In Step 2 the analyst defines the events or *scenarios* that can result in damaging consequences. The event scenarios are developed by process hazard analysis. Specifically, for scenarios involving the releases of flammable material one defines hole sizes, process conditions, locations, and ambient conditions.

In Steps 3 through 6 the *consequences* of these scenarios are evaluated. *Flammable mass* calculations are needed in Step 4 of this process. Explosion and fire calculations are needed in Step 5. Step 6 requires modeling the response of buildings to explosions, or of people to thermal radiation, flying objects, blast shock waves and the like. Steps 4 through 6 in Figure 2.1 have been further detailed by Kletz [1977] to a set of tasks as follows:

1. For each release scenario, estimate how much of the inventory in a plant is released before ignition occurs.
2. Estimate release conditions. Find expansion velocity and what fraction of the release will flash or condense (find liquid and vapor mass fractions).
3. Find how far the vapor cloud drifts before it ignites, or more generally, estimate the probability that it will ignite as a function of distance traveled.
4. Find the mass of material in the flammable range as a function of distance and time.
5. If ignition occurs, what is the probability that there will be an explosion and not just a flash fire?
6. What fraction of the flammable mass will explode and what fraction of the energy released will contribute to a shock wave?

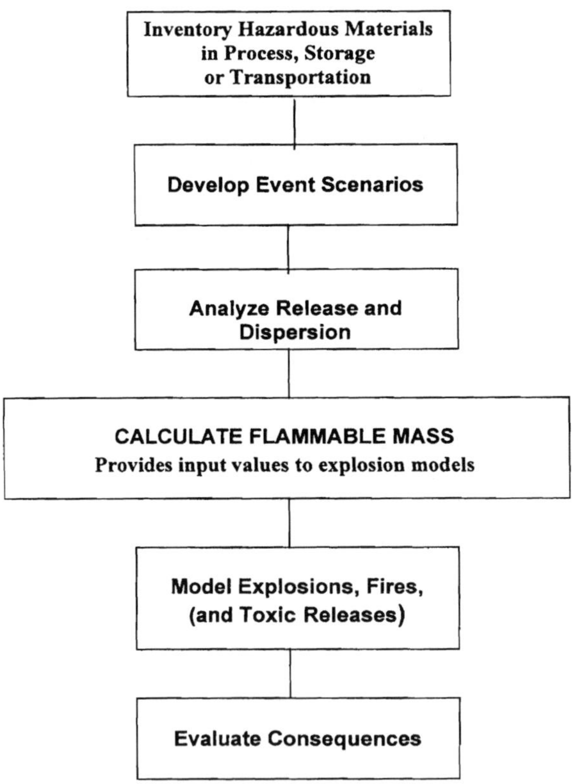

Figure 2.1. Risk evaluation process.

7. What are the effects on people and equipment for each scenario analyzed?

This book deals primarily with Tasks 3 and 4 in the above list. Secondarily, it deals with Task 2.

2.2. Objectives

The purpose of this book is to assist practicing engineers, designers, and managers of chemical facilities to realistically estimate the fraction of a cloud of accidentally released flammable material that is capable of being involved in a flash fire or explosion. It also summarizes methods to calculate blast effects from the explosion of a flammable cloud outside a building. The book pulls together in one place the pertinent and useful information that allows the reader to calculate flammable mass and explosion effects in a reliable and realistic way. Some steps can be done without a computer program or to verify a computer program. For these steps, equations are listed without derivation but with example calculations. Other steps are impractical without a computer program. This book provides our best guidance based on knowledge to date, and acknowledges what is not yet well known.

An objective is to concisely lay out methods to calculate the characteristics of a flammable vapor cloud needed as inputs to fire and explosion models. These include

- Discharge rate and phase mass fractions
- Dispersion and turbulence
- Flammability properties
- Ignition probability
- Degree of congestion and confinement
- Mass fraction of flammable cloud affected by congested/confined portions of the plant.

This book is intended to complement other CCPS Guidelines and Concept Series books which further expand the subjects of dispersion modeling, fire and explosion development and effects, hazard identification and quantification, and risk assessment. We refer to these books when appropriate and amplify a specific topic touched on in each.

This book does not treat dust explosions, pressurized vessel explosions, condensed-phase explosions, rapid phase-change explosions, boiling liquid expanding vapor explosions (BLEVEs), or runaway reaction and decomposition of reactive chemicals.

Flammable mass estimates may be needed either for hazard analysis or as part of an accident investigation. This book details predictive methods that require calculating discharge rates and amounts, air entrainment, concentration profiles, and flammable limits. Post-event methods use damage analysis from an accident and models of damage consequences to infer the flammable mass and energy required to inflict observed effects.

The book provides a number of methodologies and help in selecting the appropriate method for a desired degree of resolution. It also identifies areas where information is lacking, and recommends improvements. Another objective is to reference newer and possibly better methods not yet reduced to common practice. For example, turbulence is important in finding flame speed and overpressures, but is not yet modeled well. Models that provide only mean values of concentration and temperature are more readily available and more commonly used in practice.

1.3. How to Use This Book

This book is designed to be read through on first reading and as a workbook thereafter. The background section is in Chapters 3 and 4. The workbook portion is in Chapters 5, 6, and 7. Chapter 5 provides methods for determining flammable mass, including rules of thumb, analytic solutions, numerical solutions, and post-event damage analysis tables. Chapter 6 provides an overview of computer programs that are used to calculate flammable mass. Chapter 7 presents some worked examples.

Chapter 3 provides motivation for further study by reviewing a few examples of industry experience with fires and explosions. These illustrate the importance of calculating flammable mass as part of planning studies and risk analysis and as part of accident investigations.

The user may want to select among the wide-ranging menu of background topics covered in Chapter 4. Chapter 4 provides usable methods to calculate discharge rates, aerosol drop size, rainout, and reevaporation to generate the source rate for a vapor cloud. It also reviews the principle factors affecting dispersion and ignition of the cloud. This discussion includes how turbulence affects ignition probability and flame acceleration. To put the calculation of flammable mass in context, this chapter reviews concepts of flammability and explosions treated in depth elsewhere: flammability limits, ignition energy, autoignition temperature, flash point, flame speed. For completeness, explosion models are included here, along with a definition of all the parameters that must be specified to use them.

Chapter 8 recommends future research.

There are numerous tools available to estimate flammable mass; choosing which to use depends mostly on the cost of the methods and the justification for employing resources. At the screening level, quick estimates are needed, for example, of whether impacts from explosions and fires are likely to be felt offsite. A recommended screening level tool described here is *Dow's Fire and Explosion Index Hazard Classification Guide* (AIChE, 1994).

For many applications, a computer program may be needed. Computer programs vary from simple codified rules of thumb to complex codes which require expert assistance. The engineer needs to know the bases for such programs, and the limitations of each approach. This book indicates the present state of the art and what a user may expect from available programs and approaches. Chapter 6 lists some widely available computer programs.

Some applications require the highest accuracy, even if it means accepting higher costs to achieve that accuracy. For example, to design blast walls and escape routes on offshore platforms, engineers have employed wind-tunnel models and computational fluid dynamics (CFD) modeling.

At each level it is important to know the accuracy of available methods. Ideally, each method should be compared with experimental data. Unfortunately, good quality experimental data are in short supply. As a "poor man's" substitute for experimental validation, it is sometimes necessary to conduct sensitivity analyses, which at least show how variations in important variables affect the predictions. Chapters 4 and 5 show how to find the factors required by the methodologies.

3

INDUSTRY EXPERIENCES WITH FLAMMABLE VAPOR CLOUDS

It is the success of engineering which holds back the growth of engineering knowledge, and its failures which provide the seeds for its future development.
—Blockley and Henderson, 1980

This chapter provides examples of major incidents that have occurred releasing flammable materials. Incidents in which flammable mass estimates are reported are included in Section 3.2.

3.1. Property Losses from Vapor Cloud Accidents

Vapor cloud explosions and fires are responsible for most of the largest property loss events worldwide in the hydrocarbon industry as illustrated in **Figure 3.1**. This figure analyzes the 100 largest property losses for the period 1966–1996. Vapor cloud explosions account for the largest proportion of losses both by number of events and by dollar loss (inflation adjusted).

The type of equipment involved in the largest property loss events is shown in **Figure 3.2**. Piping systems are most frequently involved, and reactor events have the largest average dollar loss. Failure of any of the listed type of equipment could result in a flammable vapor cloud.

3.2. Examples of Vapor Cloud Events

The following examples have been selected to illustrate a variety of settings, materials, and conditions that occur, to represent the situations in which flammable mass estimation could be needed. The focus of this discussion is on flammable mass. Further details of the events are provided in referenced material.

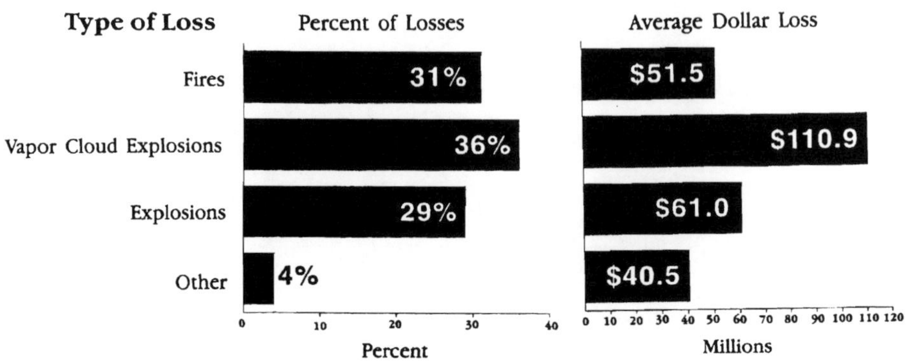

Figure 3.1. Distribution of top 100 large property damage losses in hydrocarbon processing industry by type of event, 1966–1996 (Reproduced by permission of J&H Marsh and

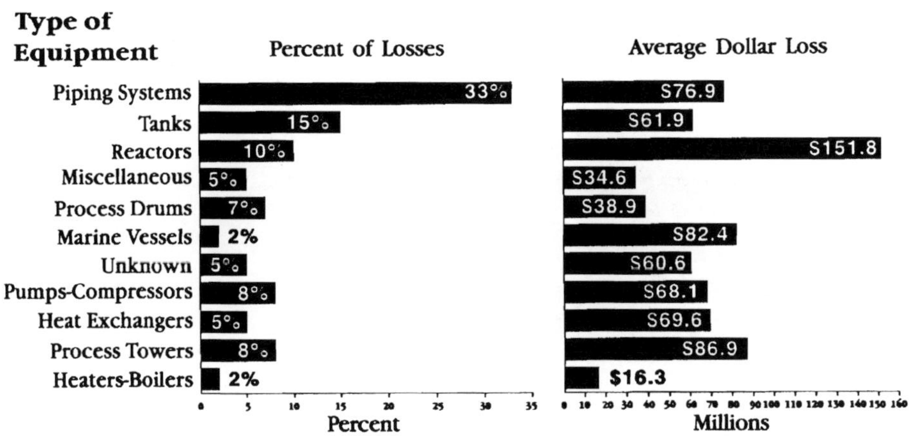

Figure 3.2. Distribution of top 100 large property damage losses in hydrocarbon processing industry by type of equipment, 1966–1996 (Reproduced by permission of J&H Marsh and McLennon.)

3.2.1. Bangkok, Thailand, LPG Vapor Cloud

At approximately 10:30 p.m. in the evening of September 24, 1990 a flatbed truck carrying two LPG tanks "recklessly and at high speed" careened off an expressway exit ramp, apparently trying to beat the traffic light at the base of the ramp [*Hazardous Cargo Bulletin,* 1990]. The truck crashed into an automobile on the six lane New Petchburi Road in the center of Bangkok (see **Figure 3.3**). The LPG tanks on the truck bed were interconnected by a 2-in. line at the bottom of the tanks. The truck tipped over immediately and the 2-in. line severed, discharging

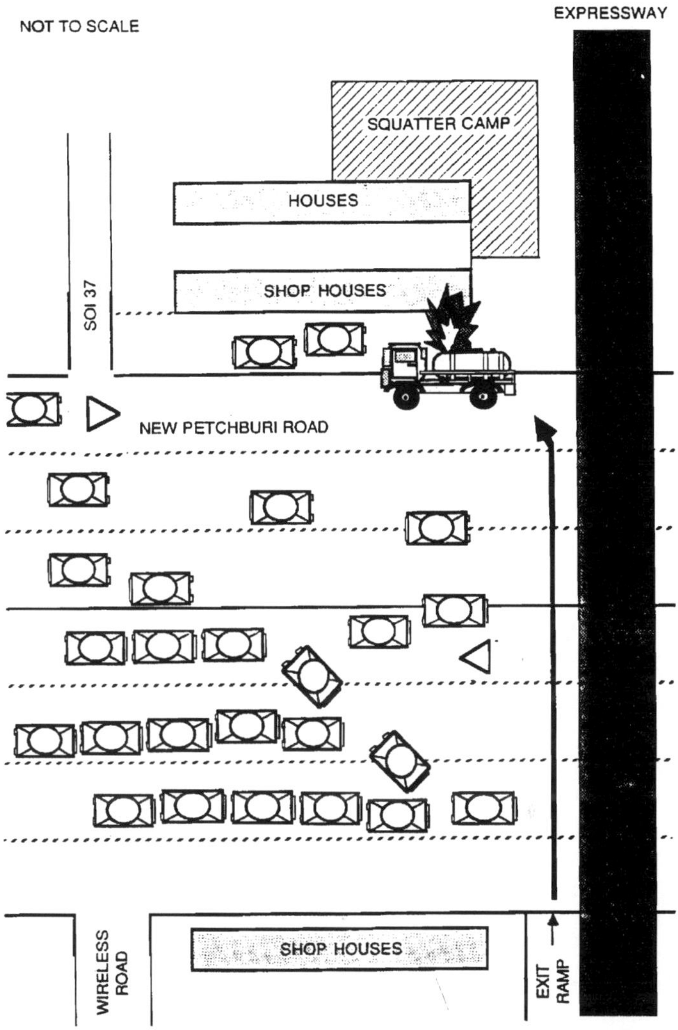

Figure 3.3. Schematic layout at the scene of the Bangkok LPG truck accident (Reproduced by permission of DNV Technica, Inc.)

some 5000 kg of LPG. The LPG formed a white vapor cloud that spread west along New Petchburi Road. The wind speed at the time was relatively calm. An eyewitness in a car near the Wireless Road intersection reported, "We could smell gas everywhere. It had filtered into the car." The cloud passed over numerous ignition sources without igniting, possibly because its concentration was above the UFL (although the edge of such a cloud always contains a flammable zone). After a considerable delay it ignited, with one report saying the flame spread east up New Petchburi Road from the corner of Soi 37. There was a flash fire, and at least one explosion, probably from gas that had entered a nearby building. Forty-eight of the shop houses on both sides of the street were destroyed, as were 57 cars. Some 68 people died and over 100 were injured. Evidently, the explosion(s) occurred in confined areas, inside shops. Flash fire occurred in the unconfined street area (Lees,1996, p. A162; Shaw, 1990).

This event illustrates how both flash fires and explosions can result from a single release and that a flammable cloud can pass over several ignition sources before igniting because not all the cloud is in the flammable range and not all ignition sources have an immediate effect.

3.2.2. Saint Herblain, France, Gasoline Cloud, October 7, 1991

A petroleum depot with a fuel storage capacity of approximately 80,000 m^3 at Saint Herblain, France adjoined a parking lot used by numerous petroleum trucks. At approximately 4 a.m. the daily activities were just starting and the atmospheric conditions were stable (Pasquill Stability Class E or F) with wind speeds less than 1 m/s. The temperature was 15 C with nearly 100% relative humidity. A leak occurred on a transfer line. Gasoline leaked continuously and developed a large aerosol cloud of about 23,000 m^3 that covered the road and parking lot and part of the storage area. Four tank trucks in the area of the cloud had their motors running. About 20 min later, the aerosol cloud was ignited and a gas explosion developed. Storage tanks were damaged and tank trucks were overturned and burned.

This event was extensively analyzed to see why the conditions led to flame acceleration and an explosion. The computational fluid dynamics (CFD) model REAGAS was set up with six parallel tank trucks inside the vapor cloud, as shown in **Figure 3.4** (Lechaudel and Mouilleau, 1995). The flame progression is predicted as shown in **Figure 3.5** for the plan view and end view. Interestingly, the simulation showed that it required at least six trucks to develop enough flame acceleration to produce explosive overpressures. **Figure 3.6** shows a large increase to damaging levels of 100 kPa when the flame passed under the sixth tank truck. When the simulation is repeated using only five trucks in the vapor cloud, the peak overpressures are predicted to be only around 10 kPa, not enough to cause serious structural damage.

The importance of this example is that it confirms findings from small-scale experiments showing that repeated plant structural elements can cause flame acceleration to explosion velocities. The Saint Herblain incident indicates that a small difference in the number of parallel-parked trucks is enough to develop a critical degree of partial congestion leading to an explosion. It also is an event

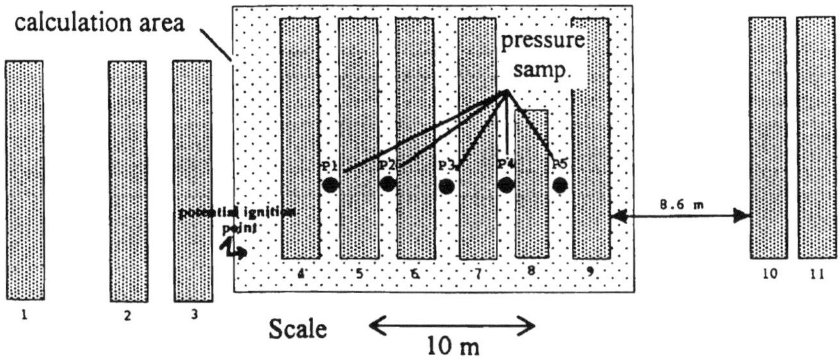

Figure 3.4. Arrangement of tank trucks inside vapor cloud as modeled by REAGAS model. (Reproduced by permission of INERIS Parc Technologique Alata.)

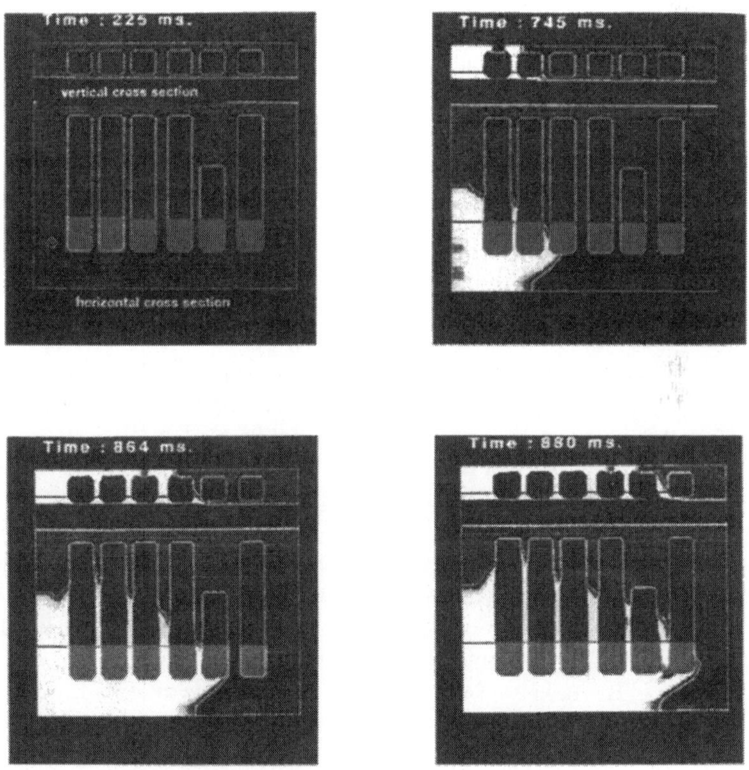

Figure 3.5. Temperature fields predicted by REAGAS model near tank trucks at Saint Herblains, France. Light color is >2500 K. dark is <1750 K. (Reproduced by permission of INERIS Parc Technologique Alata.)

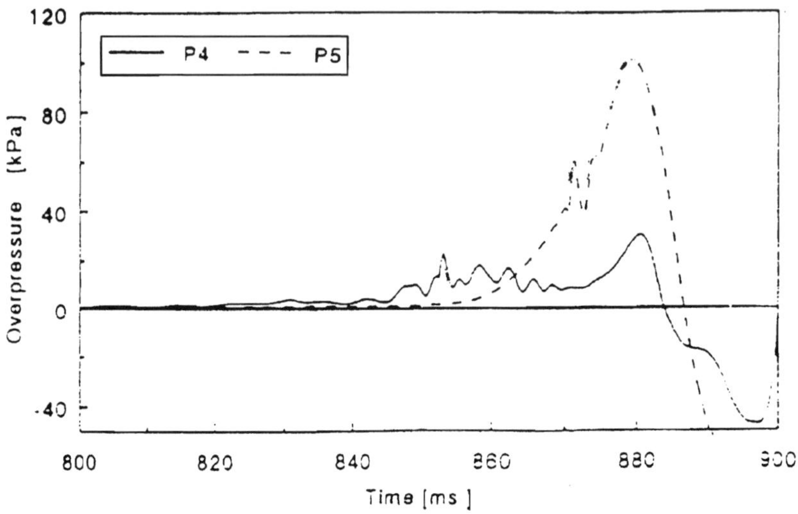

Figure 3.6. Overpressures predicted between the last two tank trucks in the vapor cloud at Saint Herblains, France. See Figure 3.4 for location of points P4 and P5. (Reproduced by permission of INERIS Parc Technologique Alata.)

occurring under meteorological conditions giving poor dispersion (F stability, < 1 m/s wind speed).

3.2.3. Pampa, Texas, Hoechst-Celanese Explosion, November 17, 1987

The Pampa plant used liquid phase oxidation of butane with air about 49 bar(a) (700 psig) to produce acetic acid and acetic anhydride. The plant was shut down because of a boiler outage, but each reactor contained about 38.0 m^3 (10,000 gallons) of reactants in the reactor and 26.6 m^3 (7000 gallons) in the decanter. Upon starting up one of the three 38 m^3 reactors, a flammable mixture in a reactor manifold reached hot catalyst and exploded, breaking the reactor feed line. Butane, acetic acid and methyl ethyl ketone discharged through the broken line and formed a vapor cloud for about 30–45 s. It was ignited by gas-fired boilers about 60 m away. The resulting explosion extensively damaged the plant and broke windows in the town of Pampa. Three people were killed and 37 injured. It also ruptured the underground fire water main, rendering the pressure inadequate to fight the subsequent fire (Mahoney, 1990; Lees, 1996, p. A1/59).

Flammable-mass modeling was used in the accident analysis to estimate the energy of the explosion and to set design parameters for blast-resistant buildings for the rebuilt plant.

3.2.4. Monsanto Ethanol Explosion, Autumn, 1970

Ethanol is considered one of the less dangerous flammable materials to handle, yet an explosion of ethanol vapors destroyed a large building. A brick building at Monsanto's Queeny, Missouri plant was 18 m (60 ft.) wide, 43 m (140 ft.) long, and 10 m (34 ft.) high with a flat roof constructed of hollow clay tiles. Originally the building had numerous large windows containing multiple glass panes, but these had all been removed and the areas left open (with 496 m^2 area). Boiling liquid ethanol overflowed from a vessel opening that was flush with the mezzanine level, intermediate between ground level and the second story. Most of the estimated 0.23–0.57 m^3 (60–150 gallons, 167–417 kg) released flowed over the edge of the mezzanine and through the open grating floor to the ground level. It cascaded over piping, reaction vessels, and structural steel on the way, vaporizing a sizable fraction.

Upon ignition there was a weak explosion, sufficient to blow out the north wall, some 6 m (20 ft.) from the spill, but not the south wall, which was cracked and had to be replaced. Roof tiles were blown upward and fell back essentially vertically. Six-meter (20 ft.) sections of the side walls were also blown out. It appeared that the pressure developed was just sufficient to exceed the strength of the building, which was estimated to fail at 69 mbar (1 psi). An explosion developed because the vapor cloud was confined and venting was inadequate. The venting provided by the open window area partially, but inadequately reduced the explosion effects. In fact, this incident provides a datum point for explosion venting design.

3.2.5. Mexico City Vapor Cloud and Explosion, November 19, 1984

When PEMEX originally constructed their LPG storage and handling facility in San Juan Ixhuatepec, or San Juanico, a suburb of Mexico City in 1962, there were no buildings or residents near the site. At the time of the incident approximately 100,000 people lived in the surrounding hills. The nearest residential building was only about 130 m south of the storage tanks. Around 5:30 a.m. on November 19, 1984, an 8″ feed line delivering LP gas (20% butane, 80% propane) ruptured inside the PEMEX tank area. A 2-m-high vapor cloud spread over an area of 150 × 200 m, as shown in Figure 3.7. There were no gas detectors, and apparently no mitigation action was taken before the vapor cloud reached an ignition source, likely the flare, and exploded at 5:45 a.m. This began a series of major "knock-on" events (Pieterson, 1985, 1986, 1988; Skandia, 1985; Lees, 1996, p. A4/1-8).

After a flash fire burned away, a ground fire was left as well as a jet fire from the yet discharging 8″ line fracture. This jet fire was directed toward one of the two large 1600 m^3 storage spheres, number F:4. Both of these nearly empty large

Figure 3.7. Estimated area covered by vapor cloud at San Juanico, Mexico terminal shown circled. The pipe rupture occurred at Point 1 and ignited at the flare at Point 2. (Reproduced with permission of Skandia International Insurance Co. Photo: State of Mexico.)

spheres exploded and ruptured other tanks. Burning LPG flowed around and through nearby houses and ignited secondary fires in the residential quarters. Several more explosions occurred, and 12 propane "bullets" rocketed up to 1000 to 1200 m away. There were 560 fatalities and 7000 injured. A large fraction of the people indoors within 300 m of the plant perished. The total destruction of the facility occurred because the protection system was destroyed, including emergency isolation and water spray systems. Traffic chaos developed as residents sought to flee while the emergency services tried to get in.

This illustrates that knock-on events can be far more destructive than the initial vapor cloud explosion. Automatic detection and protection systems are important when the release event need last only seconds to generate such a destructive series of events.

3.2.6. Pasadena, Texas Fire and Explosion, October 23, 1989

The Phillips complex at Pasadena, near Houston, Texas produced high-density polyethylene using a loop reactor with ethylene dissolved in isobutane at 49 bar (700 psi) and elevated temperature. The dissolved ethylene polymerizes to form polyethylene particles that gradually come to rest in settling legs shown in Figure 3.8 where they are removed periodically through valves at the bottom. At the top of each of these legs is a single air-actuated ball valve which is kept open during production so that the polyethylene particles can settle into the leg.

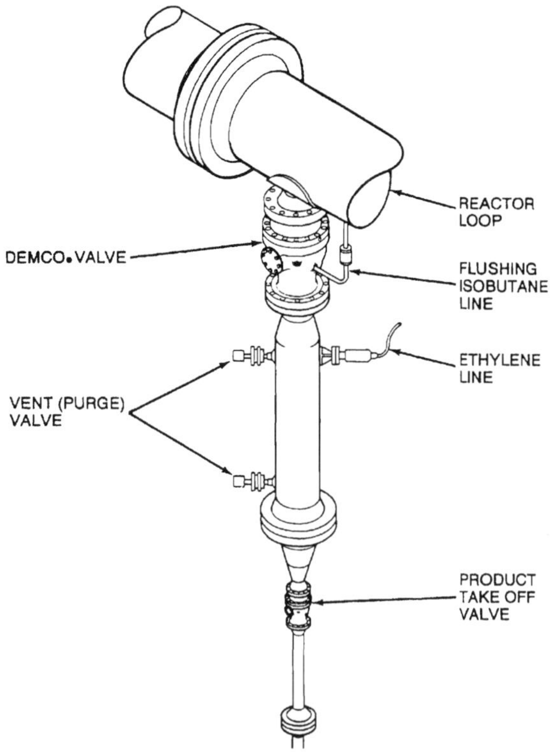

Figure 3.8. Typical settling leg on polyethylene reactor at Pasadena plant. (U.S. Dept. of Labor, A Report to the President, April 26, 1990.)

The settling legs frequently plugged with plastic material. When this happened, the ball valve was closed, the leg disassembled, and the blockage removed. If the ball valve were to open during a clean-out operation, there would be nothing to prevent the escape of flammable material. On the day of the accident, Monday, October 23, 1989, at 8:00 a.m. work began to clear the second of three plugged settling legs on Reactor No. 5. The maintenance team partially disassembled the leg and were able to remove part of the line plug, but part remained lodged in the pipe 12–18 in. below the ball valve. One of the team was sent to the control room to seek assistance. Shortly after, at 1:00 p.m. a release occurred through an open ball valve. Because of the high operating pressure, over 99% of the contents of the reactor, more than 38,550 kg (85,000 lb.), emptied in a matter of seconds.

It was subsequently established that the air hoses to the ball valve actuator had been improperly cross-connected so moving the actuator to the closed position would actually have opened it.

Reactors 4 and 5 were in a highly congested area. A huge vapor cloud, a mixture of isobutane, ethylene, hexene, and hydrogen, formed almost instantly, and was well-mixed with air by the influence of the surrounding congested piping. About 1.5 to 2 min later the cloud contacted an ignition source and exploded. A second explosion occurred 10 to 15 min later when two isobutane storage tanks exploded. Other knock-on explosions developed in a chain-reaction of explosions. One witness reported hearing ten separate explosions over a 2-hour period. Fire hydrants were sheared off by the explosion, and water pressure dropped too low to fight the fires.

Twenty-two people onsite were killed immediately, and one later died of injuries. The number injured is variously given between 130 and 300. Polyethylene plants 4 and 5 were totally destroyed. The property loss was nearly $750 million and business loss was around $650 million. Much of the damage was highly directional. For example, the administration building, located at the end of a length of pipe rack, was heavily damaged 400 m (¼ mile) from the explosion, although there was negligible damage in process areas only 30 m (100 ft) from the center of the explosion.

This illustrates the effect of congestion and confinement on an explosion as well as the highly directional nature of some explosions. The plant was rebuilt with particular attention to avoiding the high congestion of the original plant.

3.3. Examples with Postaccident Determination of Flammable Mass

The following incidents illustrate how flammable mass calculations were used in the analysis of the event.

3.3.1. Flixborough Vapor Cloud Explosion, June 1, 1974

The Flixborough Works of Nypro (UK) Ltd. was largely demolished in an explosion at about 4:53 p.m. on Saturday, June 1, 1974. This was by far the most serious accident in the chemical industry in the UK in many years, and brought about far-reaching changes in governmental oversight of plant safety throughout Europe (Lees, 1996, Appendix 2).

The plant produced caprolactam used in making nylon 6. In this process making caprolactam required making cyclohexanone by liquid-phase oxidation of cyclohexane. The oxidation was done in a train of six reactors by air injection in the presence of a catalyst at 155 C and 8.6 bar(g) (125 psig). This is above the normal boiling point of cyclohexane of 80.7 C so the pressure is some 3.5 bar(g) (51 psig) above the vapor pressure of cyclohexane [of 5.1 bar(g) (73.8 psig)]. The product from the reactors contained cyclohexanone, cyclohexanol, and 94% unreacted cyclohexane.

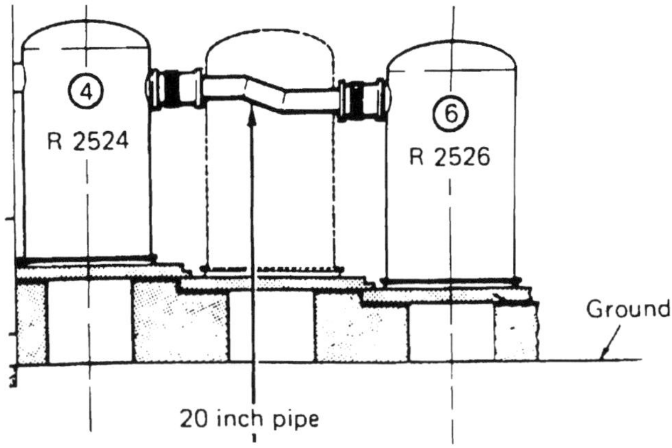

Figure 3.9. Bypass of reactor 5 at Flixborough cyclohexane oxidation plant (Parker, 1975). (Reproduced by permission of HM Stationery Office.)

On the evening of March 27, 1974 it was discovered that Reactor 5 was leaking cyclohexane and the plant was shut down. Inspection on the following morning revealed a crack some 6 ft. long in the ½″ mild steel plate. Reactor 5 was removed for repairs. Reactors 4 and 6 were connected with temporary piping so the plant could continue in production.

The reactors were normally connected by 28-inch openings with bellows for thermal expansion, but the largest pipe available on site suitable for the bypass connection was 20 in. in diameter. The two flanges were at different heights, so the connection had to take the form of a dog-leg of three lengths of 20-inch pipe welded with flanges at both ends and bolted to the existing flanges and bellows on the reactors. Figure 3.9 sketches the plant flow sheet and the 20-in. bypass. No calculations were made to check whether the bellows would withstand the twisting moments caused by the dog-leg bypass, which was not restrained in place aside from some minimal support by scaffolding.

On the morning of June 1, 1974 startup began. The sequence of events is complex and uncertain, but at a point the plant was subjected to pressures slightly higher than the normal operating pressure of 8.63 bar. During the late afternoon while the plant was on hot recycle a large quantity of cyclohexane was accidentally released. Although alternate views were put forward, the court of inquiry (Dept. of Employment, 1995) attributed the release to the rupture of the bellows and the falling away of the 20-in. bypass pipe. The resulting release formed a vapor cloud, and between 30 and 90 s after the release there was a massive vapor cloud explosion.

According to the court of inquiry report, the mass of cyclohexane discharged is estimated as 30–50 tonne from an initial inventory of 120 tonne. The

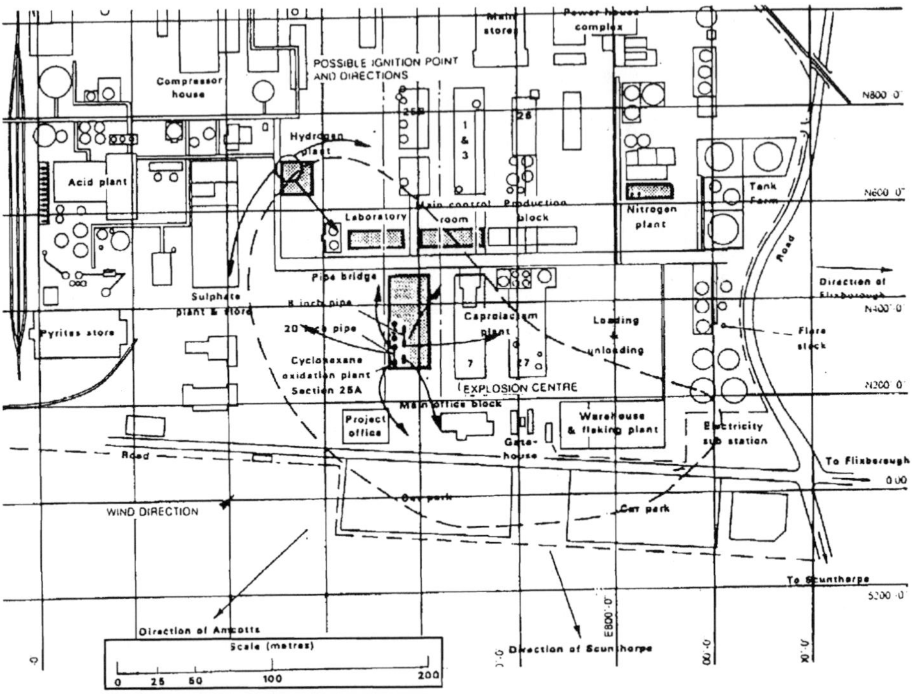

Figure 3.10 Estimated dimensions of the vapor cloud at the Nypro (UK) Ltd. Plant at Flixborough (Sadee, Samuels, and O'Brien, 1976–1977). (Reproduced by permission of Her Majesty's Stationery Office.)

estimated dimensions of the vapor cloud are shown in Figure 3.10, and formed a lenticular-shaped cloud ranging between 75 and 210 m from the source. The mechanism of failure was taken to be the simultaneous failure of both bellows and buckling of the dog-leg pipe. This failure would produce a discharge from the two 0.7-m diameter nozzels on Reactors 4 and 6. However, the court accorded this conclusion only a low probability, saying it would be readily displaced if some [other event of] greater probability could be found.

From evidence of carbon deposition and from eyewitness accounts there was a flash fire occurring at the edge of the cloud before the explosion. The cloud was expanding faster than the flame speed at first, but when the cloud expanded to a certain point, calculations show that that the expansion speed would fall below the flame speed. The flame would then travel toward the center of the cloud, analogous to a flashback from a Bunsen burner. There is evidence on seismic (ionospheric) records of two explosive events, a smaller precursor followed after 45 s by the main event (Jones & Spracklen, 1974; T. B. Jones, 1975).

An alternative detailed analysis conducted by Venart (1999) and backed by numerous eye witnesses is that only one bellows ruptured perhaps because of work hardening from flow-induced oscillations of 2–4 Hz. This is deduced from hydraulic jumps likely in the new dog-leg flow path and from calculating the natural frequency of the piping. The upstream R4 bellows was most susceptible to oscillations since it was unsupported and only partially full of fluid. When this bellows ruptured, the pipe attached to Reactor 6 was kinked back and upward in a "gooseneck" and the resulting crimp in the pipe largely prevented discharge from Reactor 6. According to this theory, the main discharge was from Reactor 4, and perhaps only 16 tonnes discharged. The flammable vapor cloud predicted by a computational fluid dynamics model depicted in Figure 3.11 envelopes the administration building in the upper corner of the figure. An indoor ignition in this building is believed to have resulted in an early explosion that destroyed the administration building. It also acted as a high-energy ignition source triggering the detonation of the surrounding vapor cloud. Such events have been observed (Harris, 1983).

Of the 28 people who died on site, 18 were in the caprolactam plant control room and were cut by flying glass and crushed by the collapsing roof. Among the general public in the surrounding area there were 53 injuries; 1821 houses and 167 shops and factories were damaged.

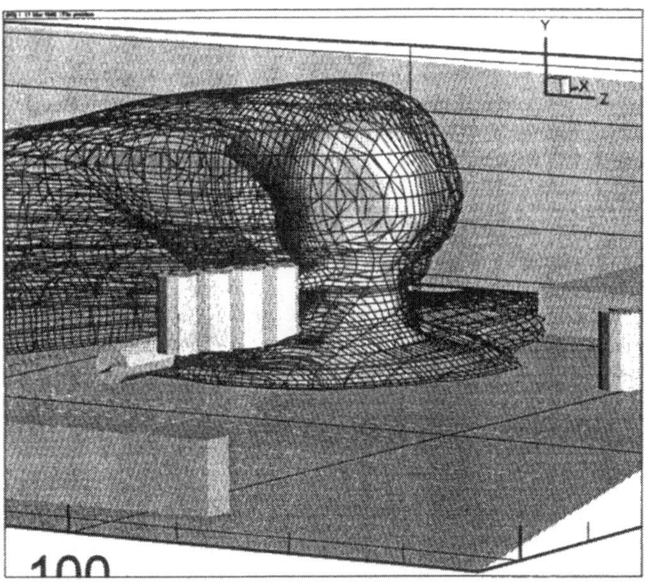

Figure 3.11. Three-dimensional plot of LFL contours of flammable vapor cloud at the Nypro (UK) Ltd. Plant at Flixborough by alternative discharge theory (Venart, 1999). Southeast view looking toward main office building (in upper corner). Discharge from R4 only at 12 s.

Estimates of explosion energy have been made, including estimates from ionospheric readings (T. B. Jones, 1975) and from the barograph of a glider in the vicinity (Gugan, 1975). The most detailed estimates are derived from damage surveys by Gugan (1976), Munday (1975), and Sadee, Samuels, and O'Brien (1976–1977). Several estimates are based on the TNT equivalence model, which range 15–45 tonne of TNT. Sadee, Samuels, and O'Brien use an alternative two-parameter model in which the height of the explosion is the second parameter. They estimate the explosion was equivalent to 16 ± 2 tonne of TNT at a height of 45 ± 24 m. Using 4.648 MJ/kg for the energy of TNT and 43.44 MJ/kg for the heat of combustion of cyclohexane, for a release of 30 tonne of cyclohexane, the gross explosion efficiency would be 5.7% based on the TNT equivalence model (Section 4.8.3). Marshall (1987) plots both near and far field damage effects and finds wide scatter in the results.

This accident illustrates that high-momentum releases can generate a vapor cloud that flows upwind. It also illustrates the possibility of an indoor detonation providing a high-energy ignition source for the outdoor vapor cloud.

3.3.2. Piper Alpha North Sea Platform Fire, July 6, 1988

The Piper Alpha platform was destroyed and 167 died when an initial explosion escalated into fires that caused the main risers to rupture. This incident was the worst accident that has occurred on an offshore platform. The inquiry that followed, presided over by Lord Cullen, produced *The Public Inquiry into the Piper Alpha Disaster (the Piper Alpha Report or Cullen Report,* Cullen, 1990), the most comprehensive inquiry conducted in the UK into any process industry disaster, onshore or offshore (Lees, 1996, pp A19/1–16).

The platform was a 10-legged steel jacket in 140 m of water. The jacket top was 62 m × 33 m. It had 36 wells, of which 26 were producing. The platform was designed with fire protection, but not with explosion protection. It was a congested, multilevel platform.

The release is attributed to miscommunication that left an open flange after removing the pressure relief valve to a condensate injection pump. The pump was isolated only by closing the gas-operated valves on the suction and discharge sides. About 9:50 p.m. the lead maintenance man organized electricians to deisolate this pump, but it is not clear what action they took. About 9:55 p.m. on July 16, 1988 signals came up in the main control room for the tripping of two centrifugal compressors in Module C. Then a third compressor tripped, followed by a high gas alarm. Personnel in the adjoining D module heard a loud screeching sound which lasted about 30 s, and then the first explosion occurred.

The inquiry concluded, using modeling, that the explosion had been caused by ignition of a gas cloud containing about 45 kg of light hydrocarbon within the flammable range arising from a two-stage leak. In the first stage, the leak rate

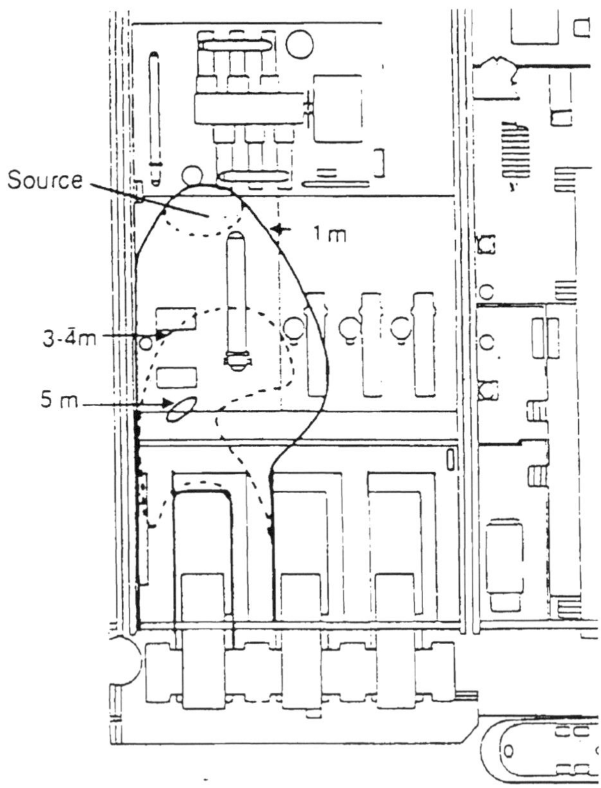

Figure 3.12. Contours of flammable gas cloud predicted by BMT wind tunnel tests: LFL contours at 30 s for leak rate of 100 kg/min. (Reproduced by permission of Controller, Her Majesty's Stationery Office.)

was probably 4 kg/s, and in the second stage some 1.83 kg/s leaked through an orifice of equivalent diameter of some 8 mm for 30 s.

Figure 3.12 portrays that the flammable cloud covered only a relatively small portion of the platform area. These predictions are consistent with readings from gas sensors. Explosion damage analysis showed that 45 kg involved in an explosion would produce 0.2 bar overpressures, enough to account for observed damage. Dynamic failure analysis showed the fire-walls fail at 0.1 and 0.12 bar overpressure.

The first explosion in the C Module destroyed most of the B/C and C/D fire-walls and severed the main communications. Crews on the adjacent platforms continued pumping which contributed to the later fires. There followed almost immediately a large fireball which issued from the west side of the B module and a large oil pool fire at that point. The large pool fire gave rise to a massive smoke plume that enveloped the platform and cut off escape routes from the living accommodation section. Some 68 men escaped by climbing down knotted ropes or jumping into the sea.

About 20 min from the initial explosion, the riser from the Tartan field ruptured. This resulted in a massive jet fire that destroyed much of the platform. About 10:50 p.m. the MCP-01 riser ruptured, and about 11:18 p.m. the Claymore riser ruptured. The pipe deck collapsed and the living quarters area tipped. By 12:15 a.m. on July 7 the north end of the platform had disapppeared. By the morning only the wellhead module remained standing.

Flammable mass calculations supported numerous design changes in offshore platforms. Platform designs now try to keep firewater systems and communication systems less vulnerable to an explosion in one location. Designs are made to prevent accumulation of an oil pool. Temporary safe refuge areas and escape routes are designed accounting for flammable clouds and heavy smoke using wind tunnels and computational fluid dynamics models. Platform enclosure is being limited to allow for cross-ventilation to disperse flammable clouds, and this makes use of flammable mass modeling. Better explosion models are being developed to assist in the design of blast walls.

3.3.3. DSM Naphtha Cracker, Beek, the Netherlands, 7 November 1975

The Dutch State Mines (DSM) Naphtha Cracking Plant II for producing ethylene had been out of service for some time and was being started up. At 9:48 a.m. a release occurred, estimated to be 5500 kg of hydrocarbon. The most likely cause of the escape was low temperature embrittlement at a weld in the feed drum of the depropanizer section (Ministry of Social Affairs, 1976). According to witnesses, the vapor cloud enveloped two sides of the control house, as shown in Figure 3.13 (Gugan, 1979). After about 2 min, an explosion occurred, ignited by one of the cracking furnaces, which destroyed the plant and a nearby tank farm and broke windows up to 4.5 km away. Fourteen people were killed, six of whom were in the control room, and eight in the open air outside the control room. Inside the plant, 104 people were injured, and 3 were injured outside.

The Dutch research organization TNO (in Ministry of Social Affairs, 1976) found 800 kg of hydrocarbon exploded, equivalent to 2200 kg of TNT. The net explosion efficiency on this basis is 28%. (See Section 4.8.3.) The investigators pinpointed the epicenter of the explosion as point E in Figure 3.13. It took a considerable distance from this point for the flame speed to build up to give damaging overpressures.

The Beek explosion was modeled in considerable detail using the FLACS model by van Wingerden et al. (1995). Their model produces overpressure contours that move through the vapor cloud as illustrated in Figure 3.14. The

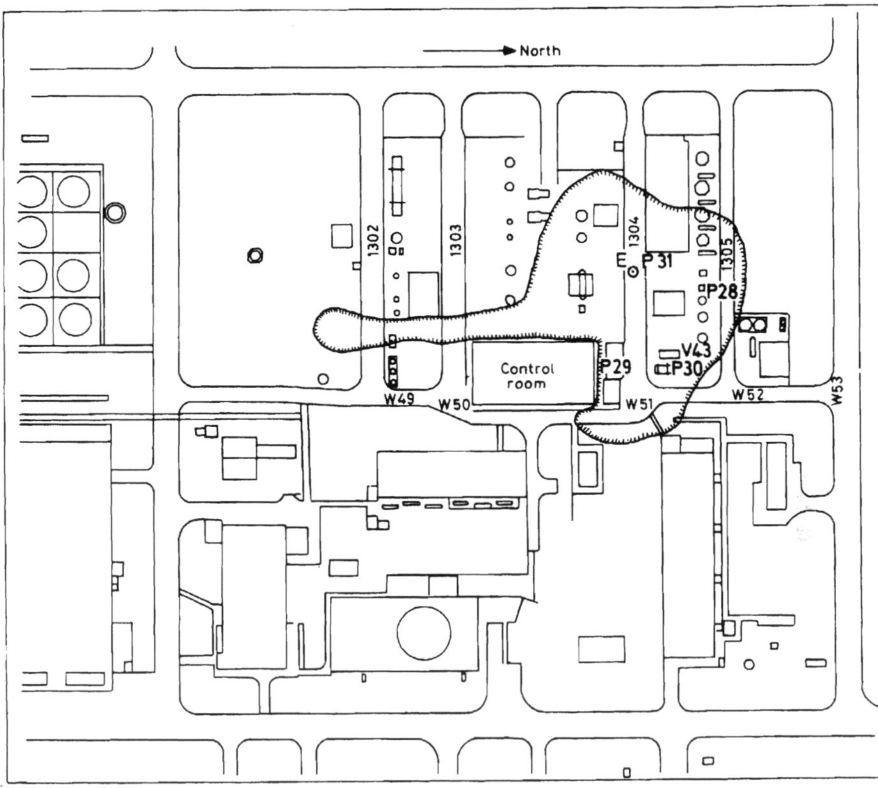

Figure 3.13. Plan view of Beek naphtha cracking plant showing vapor cloud contours reported by witnesses (Reproduced by permission of IChemE.)

influence of the control room is seen in keeping the cloud partially confined and less dilute. Their model predicts very strong explosions for an ethylene cloud if the cloud height is taken to be 3.90 to 5.73 m, but very low-pressure effects if the cloud height is reduced to 2.36 m. This illustrates an aspect of partial confinement and congestion and bears on the issue of the minimum cloud size needed for an explosion. With a vapor cloud of height <3.9 m, the mixing of air in the top of the cloud dilutes a substantial fraction of the flammable material outside of the flammable range. Furthermore, during combustion, the combustion products are readily vented out the top of the cloud and do not contribute much turbulent flame acceleration. Also, small vapor clouds may not present enough run-up distance for flame acceleration to reach critically high flame speeds.

Van Wingerden et al. also predict mild overpressures for a propylene cloud, emphasizing how the reactivity of the hydrocarbon strongly influences results. They also find a stronger explosion for a rich ethylene cloud (equivalence ratio

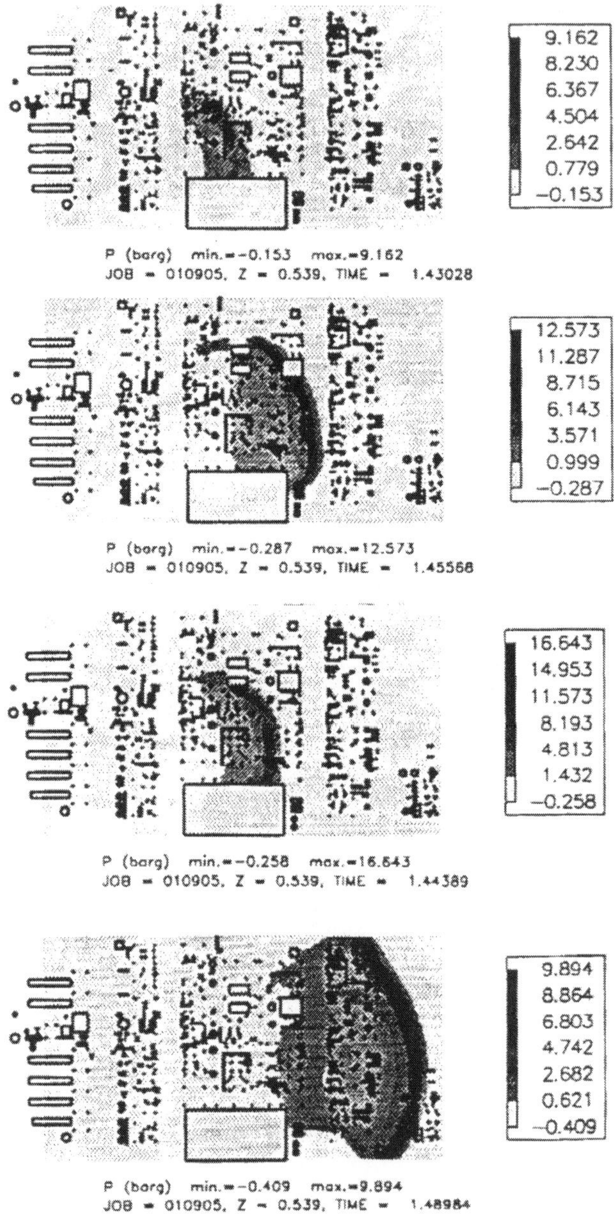

Figure 3.14. Pressure field predictions during explosion at Beek by FLACS model. Ethylene–air well-mixed at stoichiometric concentration; initial height of 5.73 m; pressure contours at 0.5 m above ground; scale at right in bar

1.5) than for a stoichiometric mixture (equivalence ratio 1.0). This prediction supports the assertion that some dilution occurs during the explosion.

Good quality flammable mass estimates for the events described above would help to improve our understanding of the energies involved, and how to design the appropriate mitigation measures. Control rooms can be moved away from processing areas or hardened. Buffer spaces can be specified between process areas and public area. Blast walls can be designed for offshore platforms to be both effective and minimal in weight. Accident analysis can better pinpoint the source of releases and of ignition. Explosion and fire models should be more successful in predicting effects or in matching observed effects if the important input values are better known. In short, the application of scientific principles to improving plant safety will be advanced.

4

BASIC CONCEPTS—FLUID FLOW, FIRES, AND EXPLOSIONS

A little fire is quickly trodden out; which being suffered,
rivers cannot quench.
—William Shakespeare, King Henry VI.

This chapter is intended to provide useable formulas to address Step 3 in the risk evaluation process flowsheet in Figure 2.1 (analyze release and dispersion). It also provides a brief background on the properties of flammable materials that affect flammable vapor clouds. The intention is not to rewrite the rich literature in combustion theory and fluid flow, but to highlight important findings. The first part of the chapter, Sections 4.1 and 4.2, discuss dispersion theory; the second part, Sections 4.3 to 4.10, discusses explosion theory.

4.1. Discharge Characteristics

Calculating the consequences of postulated accident scenarios or analyzing an accident usually begins with finding the discharge rate and the total mass discharged. The total mass discharged in an accident can often be found by measuring the inventory in the system before and after the release. However, this information alone is not sufficient to provide good estimates of flammable mass in the vapor cloud. It is preferable to calculate discharge rates, aerosol formation, rainout, cloud dilution, and concentration profiles. Thereby we find much more information such as whether the cloud disperses quickly by a high velocity jet mixing, whether it is kept heavy by evaporation of an aerosol, whether the cloud effects are prolonged by evaporation of rained-out liquid, and what fraction of the flammable portion of the cloud develops in highly congested areas.

For single-phase discharges of gases or liquids the discharge rate equations are quite simple and there are analytical solutions for time-dependent vessel blowdown. Two-phase systems are more complex.

4.1.1. Single-Phase Discharge Rates from Tanks

For subcooled liquids, that is liquids discharged below their normal boiling point, the discharge rate through a puncture or pressure relief valve depends on the **total pressure drop** across the orifice or nozzle. This is also true for gases and vapors initially at pressures below their choking pressure. **Choked flow** occurs with compressible fluids and is defined as the pressure where the discharge velocity becomes sonic (reaches the speed of sound); flow is supersonic for higher source pressures. Choked flow usually occurs when the ratio of downstream to upstream pressure is in the range of 0.5 to 0.9. Under choked conditions, the flow rate is insensitive to pressure variation **downstream** of the choke point. Discharge rates still depend on pressure **upstream** of the choke point.

SUBCOOLED LIQUID DISCHARGE

The mass flux of a general discharge is expressed in its simplest form using the dimensionless pressure ratio, η, and dimensionless mass flux, G^* defined by

$$\eta = \frac{P_{\text{out}}}{P_0}; \qquad G^* = \frac{G}{(P_0 \rho_0)^{1/2}} \tag{4.1}$$

where the upstream pressure is P_o, the downstream pressure is P_{out}, the upstream density is ρ_0, and G is the mass flux in kg/(m^{2}s). For subcooled incompressible liquid through a nozzle integration of the energy balance gives the orifice equation (the Bernoulli equation) in terms of the nozzle discharge coefficient, C_D and the expansion ratio β which is the ratio D_0/D_v of orifice diameter to vessel or pipe diameter:

$$G^* = C_D \left[\frac{2(1-\eta)}{(1-\beta^4)} \right]^{1/2} \tag{4.2}$$

or, in the original dimensional form,

$$G = C_D \left[\frac{2\rho_0 (P_0 - P_{\text{out}})}{(1-\beta^4)} \right]^{1/2} \tag{4.3}$$

For accidental releases the downstream backpressure, P_{out}, is the ambient pressure, P_{amb}. The stagnation pressure in the vessel at the elevation of the puncture, P_0, is the sum of the pressure above the liquid, P_{top}, and the gravity head (the product with the gravitational constant, g, the liquid density, ρ_L, and liquid head, h):

$$P_0 = P_{\text{top}} + g\rho_L h \tag{4.4}$$

For saturated liquids without padding gas, P_{top} is the vapor presssure, P_s. With padding gas, P_{top} equals P_s plus the partial pressure of padding gas.

The general discharge rate, w, is the product of the mass flux and the effective cross-sectional area of the puncture, A:

$$w = GA \tag{4.5}$$

In practice, liquid flow does not choke. Therefore, factors that affect the pressure downstream of the nozzle discharge will also affect the flow rate. This is the primary reason why the value of the discharge coefficient C_D for liquid flow for a smooth orifice is around 0.61, whereas values of 0.975 are more typical for C_D for gases and vapors. For two-phase flow, C_D increases from 0.61 to 0.975 as the vapor quality, x, increases from 0 to 1 (Bragg, 1960; Jobson, 1955). For ragged orifices even with liquid flow, C_D is around 0.85. A set of typical values for C_D is given in Table 4.1 (Dodge et al., 1980). The values in Table 4.1 are unconfirmed by other researchers. Table 4.1 proposes that higher values of C_D apply only with jagged petals pointing outward. The important variables to know are P_0, ρ_0, C_D, and A.

Subcooled liquid blowdown solutions are given for various tank geometries by Sommerfield (1990).

Table 4.1
Discharge Coefficients for Liquid Flow

Puncture shape	Edge condition	Orientation	C_D
Circular	Smooth	Any	0.61[a]
Circular	UPJI	Any	0.622
Circular	RPJI	Any	0.577
Rectangular	Smooth	Horizontal	0.619
Rectangular	Smooth	Vertical	0.625
Rectangular	UPJI	Horizontal	0.629
Rectangular	UPJI	Vertical	0.626
Rectangular	RPJI	Horizontal	0.609
Rectangular	RPJI	Vertical	0.613
Rectangular	UPJO	Horizontal	0.826
Rectangular	UPJO	Vertical	0.828

[a]Theoretically justified and observed value for nozzle or orifice.
UPJI = Uniform petals jagged in
UPJO = Uniform petals jagged out
RPJI = Random petals jagged in
Source: Dodge et al. (1980).

Example 1. Calculating Subcooled Liquid Discharge Rate from Orifice
Consider a horizontal propane cylinder of diameter, D_v = 4.0 m with a puncture diameter, D_0 of 0.0381 m (1.5 in.) at a depth, h, of 1.5 m below the liquid surface. With an ambient pressure of 1.01325 bar and temperature of 25 C (298.15 K) the vapor pressure of propane is 9.51 bar(a),the liquid density, ρ_L is 493.3 kg/m^3. Converting to Pascals, the stagnation pressure, by Eq. (4.4) is

$$P_0 = 9.51 \times 10^5 + 9.815 * 493.3 * 1.5 = 9.583 \times 10^5 \text{ Pa}$$

The initial pressure ratio, η is 1.01325/9.583 = 0.1057. The expansion ratio, β, is 0.0381/4. = 0.0095, so β^4 is 8×10^{-9} which is negligible. Equation (4.2) with $C_D = 0.61$ gives $G^* = 0.8158$ (typically the dimensionless flux values are of order unity). By Eq. 4.1 the mass flux is

$$G = G^*(P_{0}\rho_{0})^{1/2} = 0.8158 * (9.583 \times 10^5 * 493.3)^{1/2} = 17{,}737 \text{ kg/(m}^2\text{s)}$$

The discharge area, A is 0.00114 m^2, so the initial discharge rate is

$$w = 17{,}737 * 0.00114 = 20.2 \text{ kg/s}$$

The discharge becomes two-phase at the choke-point as the liquid head pressure decreases, but the subcooled liquid discharge rate applies initially.

Gas Discharge
The ideal gas equation of state for adiabatic, reversible (isentropic) expansion, is Pv^γ = constant where γ is the heat capacity ratio $C_p/C_v = C_p/(C_p - R/M)$ and v = $1/\rho$ is the specific volume. The ideal gas constant, R is 8314.32 J/(kgmole K) and M is the molecular weight. As γ tends toward 1, the isothermal solution results. Thus, for specific systems in which there is some heat exchange, γ is replaced by a slightly lower value, k, the polytropic constant. Thus, γ is a property of the gas; k is experimentally determined for a particular system. Values for γ are shown in Figure 4.1 for light hydrocarbon gases (Lapina, 1982). Integrating the energy balance with this equation of state gives

$$G^{*2} = C_D^2\left[\frac{2\gamma}{\gamma-1}\right]\eta^{2/\gamma}\left[1-\eta^{(\gamma-1)/\gamma}\right] \tag{4.6}$$

Choked flow occurs when the backpressure drops below a critical value, P_c, or the pressure ratio drops below η_c given by

$$\eta_c = \frac{P_c}{P_0} = \left[\frac{2}{\gamma+1}\right]^{\gamma/(\gamma-1)} \tag{4.7}$$

Substituting η_c into the more general unchoked solution, Eq. (4.6), gives for choked discharge rates

$$G_c^{*2} = C_D^2\gamma\left[\frac{2}{\gamma+1}\right]^{(\gamma+1)/(\gamma-1)} \tag{4.8}$$

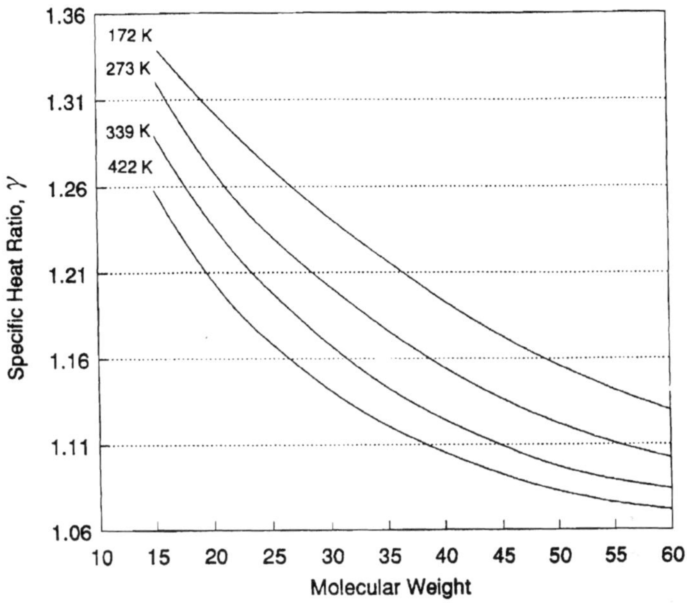

Figure 4.1. Specific heat ratio for light hydrocarbon gases as a function of molecular weight and temperature. (Reproduced by permission of McGraw-Hill, Inc.)

which applies when $\eta_{amb} = P_{amb}/P_0 < \eta_c$.

Time-dependent blowdown solutions are given for gas systems by Woodward and Mudan (1991).

Example 2. Calculation of Gas or Vapor Discharge Rate

For the propane tank in Example 1, if the puncture is, instead, a 4-inch hole in the vapor space, $D_0 = 0.1016$ m, $A = 0.008107$ m^2. The STRAPP program of the NIST (Ely and Huber, 1990) gives $\gamma = 1.1347$, $C_p = 1705.7$ J/(kg K) and the vapor density, $\rho_0 = 20.32$ kg/m^3. The choking pressure ratio by Eq. (4.7) is

$$\eta_c = \left[\frac{2}{1.1347+1}\right]^{\frac{1.1347}{1.1347-1}} = \left[\frac{2}{2.1347}\right]^{8.424} = 0.578$$

The initial pressure, P_0, is the vapor pressure 9.51 bar(a), so the initial pressure ratio η is 0.1065 which is well below the choked pressure ratio. The discharge coefficient is found, using the formulas in Appendix N, to be $C_D = 0.88$. Thus, the initial flow rate is choked and is found by Eq. (4.8) to be

$$G_c^{*2} = (0.88)^2\, 1.1347\left[\frac{2}{2.1347}\right]^{15.848} = (0.5593)^2 = 0.3128$$

Thus, from Eqs. (4.1) and (4.5):

$$G = 0.5593\ (9.51 * 10^5 * 20.32)^{1/2} = 2459\ \text{kg/(m}^2\ \text{s)}$$

$$w = GA = 2459 * 0.008107 = 19.93\ \text{kg/s}$$

Thus, a 4″ hole in the vapor space discharges propane at very nearly the same rate as a 1.5″ hole in the liquid space for this particular example. The important variables are P_0, k, C_D, and A.

4.1.2. Single-Phase Discharge Rates from Pipes

When the break or puncture occurs in a pipe, discharge rates are generally lower than when the same size puncture occurs in a vessel. This is because there are friction losses in the pipe, fittings, bends, and in pipe expansions and contractions. These effects are captured in the term N, defined by

$$N = 4f\frac{L}{D} + K_e \qquad (4.9)$$

where f is the Fanning friction factor for single-phase flow, L is the pipe length to the puncture, D is the pipe diameter, and K_e is an equivalent number of velocity heads represented by types of fittings, bends, expansions, and contractions. If the puncture point is above or below the source point (where the pipe connects to a tank, or at a pump) then a straight line drawn from the discharge point to the source point defines an angle of inclination, θ, as illustrated in Figure 4.2. A pipe inclination factor is defined for general pipe discharge cases by

$$F_I = \frac{gD\sin\theta}{4fP_0v_0} \qquad (4.10)$$

Here v_0 is the initial specific volume. F_I is positive when the discharge point is lower than the source and vice versa. When the pipe is horizontal, $\theta = 0$, and $F_I = 0$.

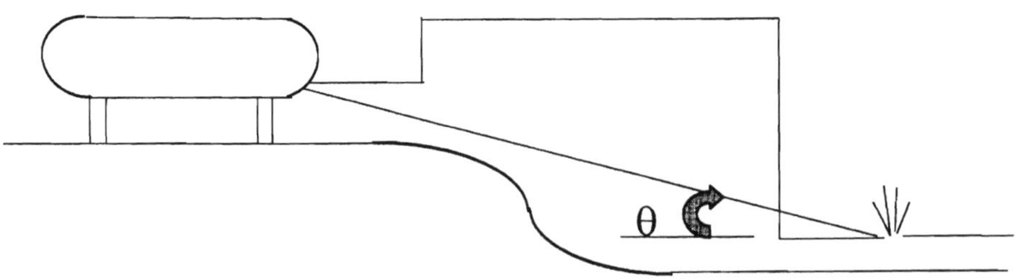

Figure 4.2. Illustration of angle of inclination for pipe break.

The orifice equation for subcooled liquids, Eq. (4.2) is modified for inclined or horizontal pipe flow to give

$$G_{\mathrm{p}}^{*} = C_{\mathrm{D}}\left[\frac{2(1-\eta)-NF_{\mathrm{I}}}{N+1}\right]^{1/2} \tag{4.11}$$

This result is strictly valid only for a complete break. For a partial break, there is a steady-state pressure drop before the break caused by the steady-state flow in the pipe. After the break, the flow upstream of the break will generally increase to account for the lowered pipe resistance caused by the break. Across a pump, this increased flow follows a pump head curve. Such complexities are best handled with either a computer program or careful attention to the specifics of the piping design.

Example 3. Calculation of Subcooled Liquid Discharge Rate from Pipe
Consider the propane tank in Example 1 with a 2-inch Schedule 40 (inner diameter = 2.067″) line running without a pump 20 m to a break point. There are five elbows and a diaphram valve three-quarters open in the path. In the first case, the line is horizontal. From friction factor plots (Perry's, 1997, pp. 6-10, 6-18), $4f = 0.014$ with a typical pipe roughness, $\varepsilon/D = 2 \times 10^{-4}$. The equivalent velocity heads for an elbows and a diaphram valve are 0.75 and 2.6, and L/D is 381, so

$$N = 0.014 * 381 + 5 * 0.75 + 2.6 = 11.7$$

Equation (4.11) gives, with $C_{\mathrm{D}} = 0.61$ and $\eta = 0.1057$ as in the example for subcooled liquid orifice flow $G_{\mathrm{p}}^{*} = 0.2291$ (horizontal pipe). The dimensional term is still $(9.583 * 10^5 * 493.3)^{1/2}$ so $G_{\mathrm{p}} = 4981$ kg/m²s. Multiplying by the area of 0.002165 m² gives the mass discharge rate of 10.8 kg/s.

In the second case, there is a 3-m elevation change. The pipe inclination factor, with $\sin\theta = 0.148$ from Eq. (4.10) is

$$F_{\mathrm{I}} = \frac{(9.184)(0.0525)(0.148)}{(0.014)(9.583 \times 10^{5})(493.3^{-1})} = 2.83 \times 10^{-3}$$

and $G_{\mathrm{p}}^{*} = 0.2270$, $G_{\mathrm{p}} = 4935$ kg/m²s, and $w = 10.7$ kg/s. These flow rates are about half the orifice discharge rate found earlier, even though the line diameter is larger than the orifice diameter in the previous example. The important variables are L/D, K_{e}, P_0, C_{D}, A, and possibly F_{I} if the elevation change is substantial (that is, the absolute value of F_{I} is close to unity). In reality, flashing flow would occur, decreasing the discharge rate by about 20% for propane.

4.1.3. Two-Phase Discharge Rates from Tanks

There are two basic types of two-phase (gas/liquid) flow. The first is the discharge of a superheated, flashing liquid such as chlorine or ammonia at 20 C

(68 F). This produces **flashing flow**, characterized by the vapor mass fraction, x, increasing along the flow path as pressure is reduced. The second occurs when a noncondensable gas such as the padding gas (nitrogen or methane typically) is mixed into the discharging liquid, or a runaway reaction generates noncondensable gases. This produces **frozen flow**, so named because the mass fraction of vapor remains nearly constant along the flow path as the pressure decreases. The vapor fraction x is also called the void fraction or vapor quality. In a so-called hybrid system, or **hybrid flow**, both noncondensable gases and flashing vapor are present at the same time.

For two-phase discharge from a pipe of length greater than about 1 m, the assumption that phase equilibrium is achieved is justified. Furthermore, the velocity of both phases is about equal during the discharge. These are the assumptions of the homogeneous equilibrium model (HEM). Two-phase discharge from a vessel or a pipe shorter than about 1 m long do not attain phase equilibrium, and HEM models tend to underpredict discharge rates (Alix et al., 1998). Better estimates require using a nonequilibrium model (NEM) or a slip equilibrium model (SEM). The SEM allows the vapor/gas velocity to exceed the liquid velocity until drag effects equilibrate the two. These methods are summarized in Woodward (1993).

We show here how to use the HEM and a simple, semiempirical NEM model, the equilibrium rate model (ERM).

HOMOGENEOUS EQUILIBRIUM MODEL (HEM)

High pressure releases are generally highly turbulent, so with enough length, as in pipe discharges, the fluid phases can be considered well mixed and uniform. It is then justified to model as a single fluid with volume-averaged properties of the two phases. There is no slip velocity between the phases. Furthermore, the two phases are assumed in local thermodynamic equilibrium, and that flow is isentropic (no friction loss). These are the assumptions of the homogenous equilibrium model (HEM).

For two-phase discharge, the two-phase density is given in terms of the vapor quality, x, by

$$v = \frac{1}{\rho} = \frac{x}{\rho_g} + \frac{1-x}{\rho_L} \tag{4.12}$$

When the gas is noncondensable, for frozen flow, we denote the initial mass vapor quality by x_0, or the corresponding volume ratio of gases by α_0. These two quantities are related by the ratio of specific volumes at the inlet conditions of vapor, $v_{v0} = 1/\rho_g$, to the two-phase value given by Eq. (4.12):

$$\alpha_0 = x_0 \frac{v_{v0}}{v_0} \tag{4.13}$$

With flashing liquids, x is given by an entropy, S, or enthalpy, H, balance from the initial state (point 0) to the final saturation temperature (normal boiling point). The subscripts refer to the phase: L for liquid, g for gas or vapor, fg for the difference between vapor and liquid. This balance can be expressed generally for a hybrid system of noncondensables plus flashing liquids to give

$$x = \frac{S_{\mathrm{L0}} + x_0 S_{\mathrm{g0}} - S_{\mathrm{L}}}{S_{\mathrm{fg}}} \cong \frac{H_{\mathrm{L0}} + x_0 H_{\mathrm{g0}} - H_{\mathrm{L}}}{H_{\mathrm{fg}}} \tag{4.14}$$

A computer program can integrate the momentum balance numerically, using the two-phase expressions for density and vapor quality along an isentropic path from P_0 to P until the mass flux G reaches a maximum. For multicomponent mixtures this involves a series of flash fractions at successive pressures.

Calculations without a computer program require analytic solutions and plots as is shown next. To obtain an analytic solution some simplification of the two-phase relationship for density as a function of pressure is needed. Several empirical relationships were evaluated by Simpson (1995), but only one gives rise to an analytic solution, the so-called omega method developed by Leung in an extensive set of papers (1986 to 1996). This uses the following approximation:

$$\frac{\rho_0}{\rho - 1} = \omega\left(\frac{P_0}{P - 1}\right) \tag{4.15}$$

Letting $\varepsilon = v/v_0 = \rho_0/\rho$ this equation is expressed as

$$\varepsilon - 1 = \omega\left(\frac{1}{\eta} - 1\right) \tag{4.16}$$

Clearly, ω is the slope of the density/pressure curve expressed in the form of Eq. (4.15) or (4.16). To help visualize this, Figure 4.3 shows a typical plot of ρ_0/ρ or v/v_0 against P_0/P, or $1/\eta$, for ammonia at an initial pressure of 20.05 bar(a) (Woodward, 1995b). The plot of ρ_0/ρ is the initial density divided by the two-phase density obtained from an equation of state. This line is slightly curved. Equation (4.15) produces a straight line with the same initial slope as the actual density curve. However, the end point of this approximate curve at the high-pressure end (ε_2, η_2) is in considerable error. A better approximation can be obtained when an equation of state is available by using a straight line between two known end points $(\varepsilon = 1, \eta = 1)$ and (ε_2, η_2), as is also illustrated in Figure 4.3. An alternative formula for ω is

$$\omega = \frac{\varepsilon_2 - 1}{\eta_2^{-1} - 1} \tag{4.17}$$

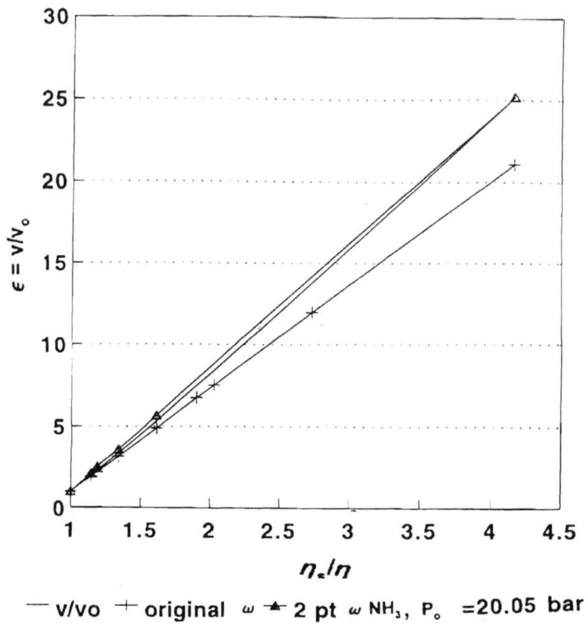

Figure 4.3. Physical meaning of the omega term in the omega method for two-phase discharge rate predictions. (Reproduced by permission of Butterworth Heinemann publishers.)

The HEM predictions for two-phase nozzle discharge are summarized in Perry's Handbook [1997, pp. 26-88 to 26-90] and shown in Figures 4.4 and 4.5. Figure 4.4 is actually two plots back-to-back, with the omega method solution for frozen flow on the left and for flashing flow on the right. The dimensionless mass flux is plotted on one set of curves; the choking pressure ratio on the other. The parameter on the curves is k, or approximately the heat capacity ratio γ. Single-phase flow is represented by the curves crossing the $\omega = 1$ axis.

To find a value of the dimensionless mass flux from Figure 4.4, all that is needed is to find ω and k. The value of ω can be found by calculating the two-phase vapor density at pressures P_0 and P_2 and then using Eq. (4.17). It can also be found using only the properties of the discharged material at the initial conditions, P_0 and T_0. Table 4.2 summarizes the omega formulas in terms of the heat of vaporization at the source point, h_{fg0}, the difference in specific volume between vapor and liquid, v_{fg0}, the liquid heat capacity, C_p, the initial two-phase density, ρ_0, and the initial volume fraction of noncondensables, α_0, if noncondensables are present.

Before calculating the two-phase discharge rate for a puncture below the liquid level in a tank, however, it is necessary to check whether the two-phase equations apply or whether the highly subcooled liquid equation applies. If the

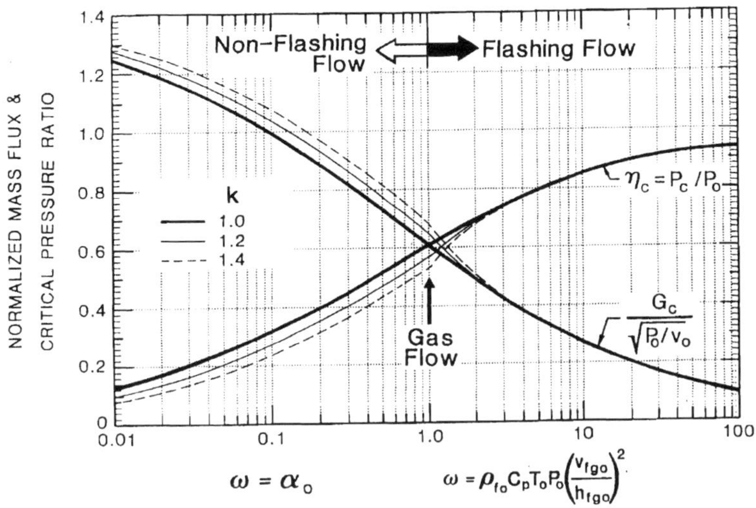

Figure 4.4. Two-phase discharge rates and choked pressure ratio for flashing and frozen flow by the homogeneous equilibrium model of Leung.

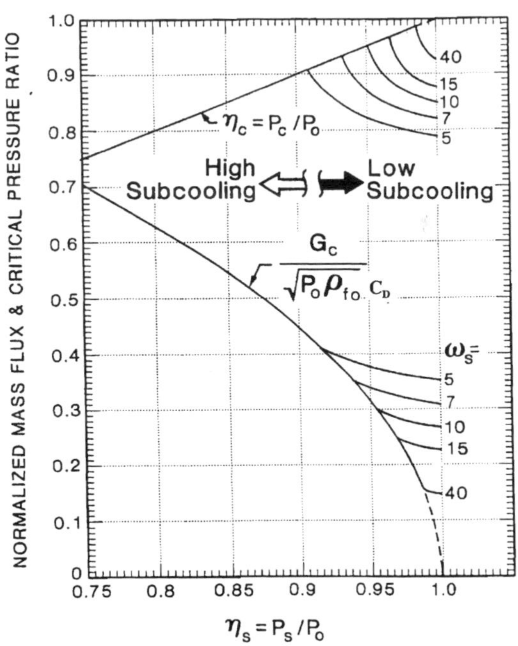

Figure 4.5. Two-phase discharge rates and choked pressure ratio as a function of the saturated pressure ratio by the homogeneous equilibrium model of Leung. This shows when two-phase flow calculations applies (high subcooling) and when subcooled liquid discharge applies (low subcooling).

Table 4.2
Formulas for Calculating Omega to Use Figure 4.2

Phase	Value of ω	Formula
All gas/vapor	$\cong 1$	
Flashing flow	$\omega > 1$	$\omega = C_p T_0 P_s \rho_0 \left(\frac{v_{fg0}}{H_{fg0}} \right)^2$
Frozen flow (nonflashing)	$0 < \omega < 1$	$\omega = \alpha_0$
Hybrid flow	$\omega > 1$	$\omega = \alpha_0 \left(1 - 2 \frac{P_0 v_{fg0}}{H_{fg0}} \right) C_p T_0 P_s \rho_0 \left(\frac{v_{fg0}}{H_{fg0}} \right)^2$
Liquid	0	

liquid is saturated at the surface, it will be slightly subcooled by the pressure of the liquid head above it. As a puncture develops, the pressure is decreased to atmospheric past the break, so at some intermediate point the pressure drops below the vapor pressure or saturation pressure, P_s, and the liquid flashes. The saturation value of η is

$$\eta_s = \frac{P_s}{P_0} \tag{4.18}$$

If the point of flashing occurs close to the puncture the discharge will be two-phase flow, and a choke point develops. The effective backpressure for the discharge will be the vapor pressure. Figure 4.5 illustrates that there are essentially two branches to the discharge rate solution, one for high subcooling and one for low-subcooling or flashing flow. By the principle of maximizing the mass flux, the applicable branch is the larger of the two. Empirically, the point at which these branches of the solution cross is given by

$$\eta_{st} = \frac{2\omega\eta_s}{2\omega\eta_s + 1} \tag{4.19}$$

When $\eta_s > \eta_{st}$ the material is at low subcooling, flashing occurs before the choke point, so the two-phase discharge rate should be found from Figures 4.4 and/or 4.5. Otherwise, when $\eta_s < \eta_{st}$ the material is highly subcooled, no flashing occurs before the incipient choke point, so use the orifice equation, Eq. (4.2).

Example 4 Calculation of Flashing Two-Phase Discharge Rate
Consider a tank of saturated liquid monomethylamine (MMA) at 37 C (98.6 F, 310.15 K) with 0.75 m of liquid head above a 2-in. (0.0508-m) puncture. The

physical properties of MMA by the DIPPR system (Daubert and Danner, 1989) are as follows:

Variable	Value	Units
Molecular weight, M	31.057	kg/kmol
Normal boiling point, T_b	266.82	K
Vapor pressure, P_s and top pressure, P_0	5.268	bar(a)
Liquid density, ρ_{L0}	638.7	kg/m^3
Vapor density, ρ_{v0}	7.087	kg/m^3
Liquid heat capacity, C_{PL}	2994	(J/kgK)
Liquid entropy, S_{L0} at T_0	–2783	(J/kgK)
Heat of vaporization, h_{fg}	724.7×10^3	(J/kg)
$C_p/C_v = \gamma$	1.190	(dimensionless)

Figure 4.6 illustrates this example. The vapor pressure, P_s, is used instead of the source pressure, P_0, because flashing starts when the pressure decreases to P_s. From Eqs. (4.4):

$$P_0 = P_s + g\rho h = 5.268 + (9.814)(638.7)(0.75)1 \times 10^{-5} = 5.315 \text{ bar(a)}$$

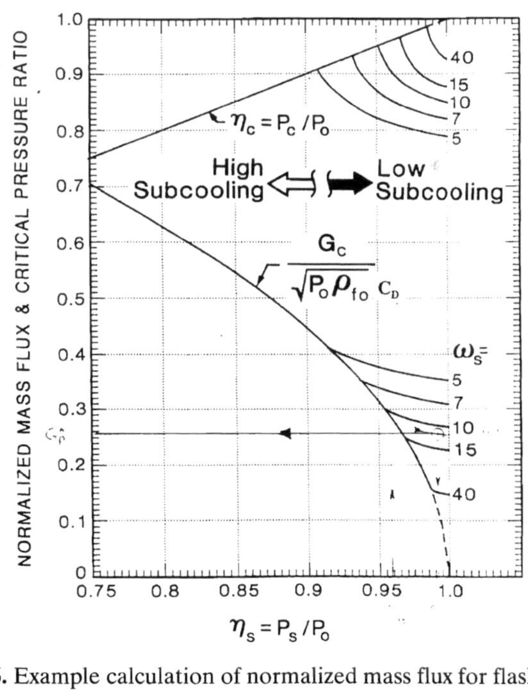

Figure 4.6. Example calculation of normalized mass flux for flashing liquid.

From Eq. (4.18),

$$\eta_s = 5.268/5.315 = 0.9912 \text{ (slightly subsaturated)}$$

$$v_{fg0} = \frac{1}{7.087} - \frac{1}{638.7} = 0.1395 \text{ m}^3/\text{kg}$$

From Eq. (4.19),

$$\omega = C_{PL} T_0 P_s \rho_0 \left[\frac{v_{fg0}}{H_{fg0}} \right]^2$$

$$= (2994)(310.15)(5.268 \times 10^5)(638.7) \left[\frac{0.1395}{724.7 \times 10^3} \right]^2 = 11.6$$

$$\eta_{st} = \frac{2(11.6)(0.9912)}{1 + 2(11.6)(0.9912)} = 0.958$$

Since $\eta_s > \eta_{st}$, look up the value on the upper branch of the lower branch of Figure 4.6 (labeled ω_s), interpolating between $\omega = 10$ and 15, to obtain $G^* = 0.256$. This gives, with $C_D = 0.938$ from Appendix M:

$$G = C_D G^* (P_0 \rho_0)^{1/2} = (0.938)(0.256)(5.315 \times 10^5 \times 638.7)^{1/2} = 4422 \text{ kg/(m}^2\text{s)}$$

$$w = GA = (4422)(0.002027) = 8.96 \text{ kg/s}$$

4.1.4. Two-Phase Discharge Rates from Pipes

It is particularly difficult to treat partial breaks from pipes since the steady-state flow and pressure profiles that exist before the break are altered by the break. Flow reversal can even occur for portions of the piping network. In the subsections below two simple models are given for two-phase discharge for the simple case of a horizontal pipe with a complete break. The first is the HEM, extended to pipe flow. The second is an approximate, but often adequate method, the equilibrium rate model (ERM).

HEM FOR HORIZONTAL PIPE FLOW

Figure 4.7 gives the HEM predictions for horizontal pipe flow as a function of the resistance factor, N. This figure is given in Leung and Grolmes (1987), Leung and Epstein (1990) and summarized in Perry's Handbook (1997, pp. 26-84 to 26-92). The ordinate is the ratio of the flow G^* for pipe flow to G^* for orifice flow (or

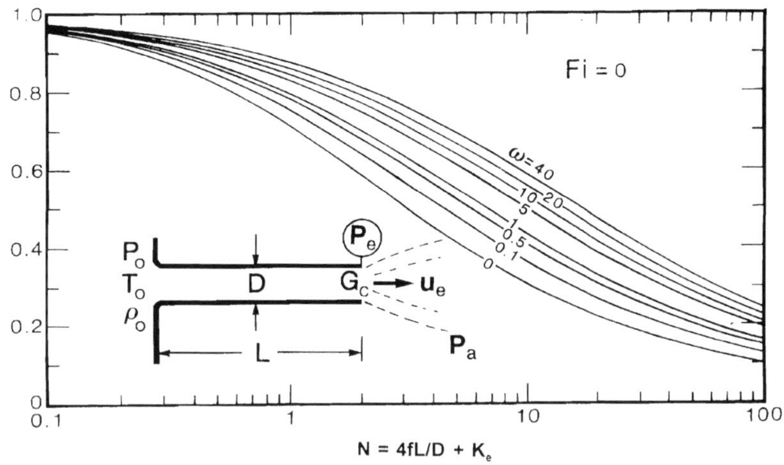

Figure 4.7. Ratio of mass flux for horizontal pipe flow to mass flux for orifice flow as predicted by the homogeneous equililbrium model.

G_p/G_{orifice}). The procedure is to first solve for G^*_{orifice} using the orifice flow charts, as above, and then apply a correction to obtain pipe flow using Figure 4.7 and

$$G^*_{\text{p}} = G^*_{\text{orifice}}\left(\frac{G^*_{\text{p}}}{G^*_{\text{orifice}}}\right) \tag{4.20}$$

Example 5. Calculation of Two-Phase Discharge from Horizontal Pipe

Using the conditions for Example 4, consider two cases in which a break occurs at the 3-m length in a 2-inch Schedule 40 pipe (0.0525 m ID): (1) without the elbows and diaphram valve; (2) with five elbows and a three-quarters open diaphram valve. Use the same friction factor as in the previous pipe example, $4f = 0.014$. The pipe resistance factor is

$$N = 4f(L/D) + K_e = 0.014\ (3/0.0525) + 0 = 0.80 \text{ without elbows}$$

$$= 0.014\ (3/0.0525) + 6.35 = 7.15 \text{ with elbows}$$

From Figure 4.7, we obtain for $\omega = 11.6$, the following results with $C_D = 1.0$:

Variable	**Case 1**	**Case 2**
N	0.80	7.15
$G^*_{\text{p}}/G^*_{\text{max}}$	0.870	0.595
G^*_{p}	0.222	0.152
G_p, kg/(m^{2}s)	4099	2800
w, kg/s	8.87	6.06

EQUILIBRIUM RATE MODEL (ERM)

A particularly simple nonequilibrium model (NEM) was suggested by Fauske (1985)and Grolmes et al. (1990) called the equilibrium rate model (ERM). This model makes an orthogonal vector addition of two rate terms:

$$G_{\text{ideal}}^2 = G_{\text{L}}^2 + G_{\text{erm}}^2 \tag{4.21}$$

The term G_{L} is the ideal subcooled liquid discharge rate (Bernoulli equation) or Eq. (4.2) with C_{D} set to 1.0.

$$G_{\text{L}}^2 = \left[\frac{2\rho_{\text{L}}\Delta P}{(1-\beta^4)}\right] \tag{4.22}$$

$$\Delta P = P_0 - P_{\text{s}}(T_0)$$

the difference between the tank pressure and the vapor pressure or saturation pressure. The second term is the equilibrium rate term:

$$G_{\text{erm}}^2 = \frac{G_{\text{CL}}^2}{N_{\text{eq}}} \tag{4.23}$$

The numerator of Eq. (4.23) is derived using the Clapyron equation in the same way that ω is derived

$$G_{\text{CL}}^2 = \frac{1}{TC_{\text{PL}}}\left[\frac{H_{\text{fg}}}{v_{\text{fg}}}\right]^2 = \frac{P_{\text{v}}\rho_0}{\omega} \tag{4.24}$$

The nonequilibrium term, N_{eq} is defined as

$$N_{\text{eq}} = 10L + \frac{G_{\text{CL}}^2}{G_{\text{L}}^2} \tag{4.25}$$

Combining and multiplying by the discharge coefficient, C_{D}, gives the working form of the ERM:

$$G = C_{\text{D}}\left[G_{\text{L}}^2 + G_{\text{erm}}^2\right]^{1/2} \tag{4.26}$$

The ERM predictions are intermediate between those of the orifice equation and the HEM as illustrated in Figure 4.8. Figure 4.8 shows the predictions of the HEM for pipe $L/D = 40$, the orifice equation for $L/D = 0$, and the ERM for $L/D = 12$. The pressure driving force term in Eq. (4.22) goes to zero for saturated liquids, but a valid value is found with a slight amount of subcooling.

Example 6 Calculation of Two-Phase Discharge by ERM

Consider a tank of water at 50 bar(a) with a broken outlet 1-in. pipe 2 m long. The saturation temperature for 50 bar is about 537.25 K. With a tank diameter of

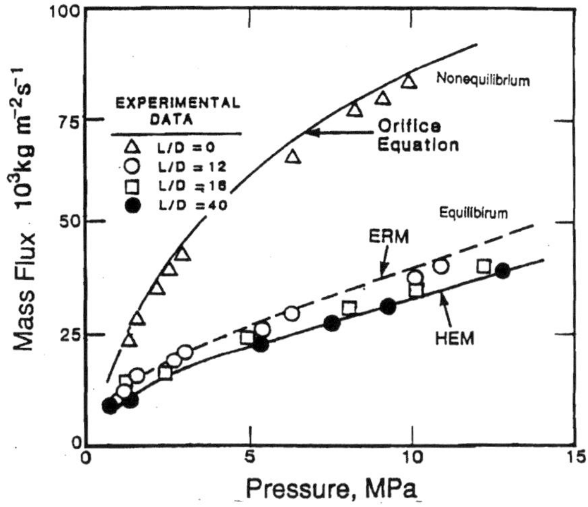

Figure 4.8. Comparison of model predictions for flashing discharge of initially saturated water with data of Fauske (1965).

3 m, $\beta = 0.0254/3 = 0.0085$, so the β^4 term is negligible. The solution is tabulated on the next page for three different degrees of subcooling, and the results plotted against experimental data in Figure 4.9. Figure 4.9 also shows the predictions for $L = 0.2$ m. The agreement with experimental data is good. Example calculations:

$$G_{CL}^2 = \frac{1}{(488.4)(4568)}\left[\frac{1.934E6}{0.03268}\right]^2 = 1.5695E9 \text{ in } (kg/m^2s)^2$$

$$G_L^2 = [2\rho_L \Delta P] = 2(928.0)(28.88E5) = 5.360E9 \text{ in } (kg/m^2s)^2$$

$$N_{eq} = 10L + \left[\frac{G_{CL}^2}{G_L^2}\right] = 20 + \frac{1.569E9}{5.360E9} = 20.293$$

$$G_{erm}^2 = \frac{G_{CL}^2}{N_{eq}} = \frac{1.569E9}{20.293} = 7.734E7 \text{ in } (kg/m^2s)^2$$

$$G = G_D[G_L^2 + G_{erm}^2]^{1/2} = 1.0[5.360E9 + 7.734E7]^{1/2} = 73{,}738 \text{ } (kg/m^2s)$$

4.1.5. Aerosol Formation and Drop Size Correlations

Aerosols are readily formed with discharges of two-phase or liquid streams or condensing vapors. A thorough treatment of aerosol theory and rainout is given in the CCPS Concept Series book *RELEASE, A Model with Data to Predict Aerosol Rainout in Accidental Releases* (1998).

Subcooling, C	–49	–29	–4
Tank temperature, T_0, K	488.4	508.4	533.4
Vapor pressure, $P(T_0)$, bar(a)	21.12	30.72	47.08
P, bar	28.88	19.28	2.92
Liquid density, kg/m^3	928.0	917.4	902.2
Vapor density, kg/m^3	29.62	27.07	24.72
C_{pL}, J/(kg K)	4568	4712	4965
h_{fg}, J/kg	1.934E6	1.854E6	1.734E6
v_{fg}, m^3/kg	0.03268	0.03585	0.03935
G^2_{CL}	1.569E9	1.164E9	0.7334E9
G^2_L	5.360E9	3.5375E9	0.5269E9
N_{eq}	20.293	20.316	21.39
G^2_{erm}	7.734E7	5.495E7	3.428E7
G, kg/(m^{2}s)	73,738	59,937	23,689
w, kg/s	37.36	30.37	12.00

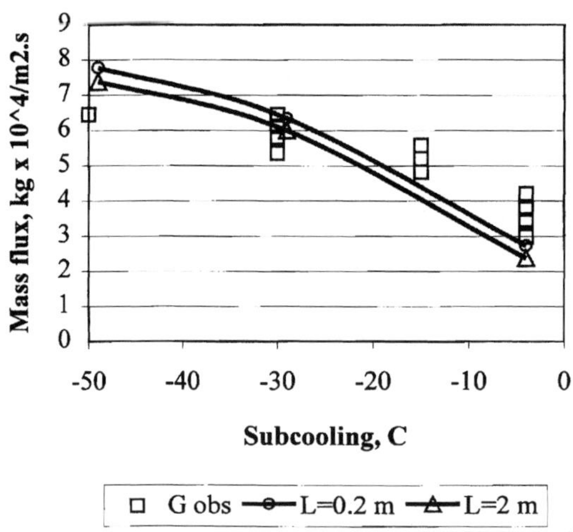

Figure 4.9. Comparison of ERM predictions with experimental data (Fauske, 1985).

Aerosols can form when hot vapors of a low-volatility material discharge into cool air. When this happens the material condenses to form a condensation fog, a relatively stable mist of small droplets. Condensation fogs typically have a near monodisperse drop size distribution, since all of the drops form essentially at the same time and grow under nearly the same conditions. Since the overall density of aerosols is higher than that of corresponding vapors, the flammable mass is much higher in aerosols than in vapors.

Aerosols also form by a pressurized discharge of subcooled or superheated liquids. Aerosol formation occurs by one of three mechanisms: (1) capillary breakup, (2) aerodynamic breakup, or (3) flashing breakup.

Capillary breakup occurs when subcooled liquids discharged through a very small hole (<2 mm diameter) form a cylindrical stream which pinches off by surface tension forces to a string of nearly monodisperse drops. The mean droplet diameter resulting from capillary breakup, d_p is proportional to the orifice diameter, D (Lienhard and Day, 1970):

$$d_p = 1.9D \tag{4.27}$$

Such pinhole leaks of flammable materials are not generally consequential if the leak is outdoors since the discharge rate is small.

Aerodynamic breakup occurs with larger punctures with subcooled liquids or slightly superheated liquids. Aerodynamic breakup occurs by mechanical shattering and peeling mechanisms. It is correlated with the Weber number, which is the ratio of shear forces on the surface of the liquid, proportional to the velocity heads term, $\frac{1}{2}\rho_g \Delta u_d^2$, to surface tension pressure, σ/d_p:

$$We = \frac{1}{2}\frac{\rho_g \Delta u_d^2 d_p}{\sigma} \tag{4.28}$$

Experimentally, droplet breakup occurs at a critical value of the Weber number, We_c, between 12 and 22 (Brown and York, 1962; Bushnell and Gooderum, 1968; Hinze, 1955). Rigorously, u_d is the relative velocity of the drop, that is the drop velocity (equal to the expansion velocity, $u_d = u_{exp}$) minus the wind speed, u_w. Since the wind speed is usually much less than the expansion velocity, only u_d is used in Eq. (4.28). The mean drop size, d_p is found by rearranging the definition of Weber number, setting $We = We_c$

$$d_p = \frac{2\sigma We_c}{\rho_g u_d^2} \tag{4.29}$$

For example, using a critical Weber number of 12 with n-octane stored at 300 K, the predicted drop size by aerodynamic breakup decreases as the tank pressure increases as shown below.

Tank Pressure (bar)	Expansion velocity (m/s)	Mean drop diameter (m)
1.0	17.5	1377
2.0	24.3	712
4.0	34.1	362
5.0	38.1	290
7.5	46.6	194
10.0	53.7	146
15.0	65.7	98

Flashing breakup occurs as **superheated** liquids are discharged and part of the liquid flashes to vapor in the form of bubbles which grow rapidly and break up the surrounding liquid. An exception is noted by Lienhard and Day (1970) who found that jet streams of small diameter cool quickly enough to obviate flashing effects. Larger drops are consequently formed from small-diameter streams.

A correlation for drop size by flashing breakup developed to fit rainout data from aerosols is shown in Figure 4.10 as a function of superheat (CCPS, 1998). Rainout tests were conducted with cyclohexane, monomethylamine, chlorine, CFC 11, and water. The correlation is given by

$$d_{\mathrm{p}} = \frac{3 \times 10^4}{10 + 4.0 \Delta T_{\mathrm{sh}}} \tag{4.30}$$

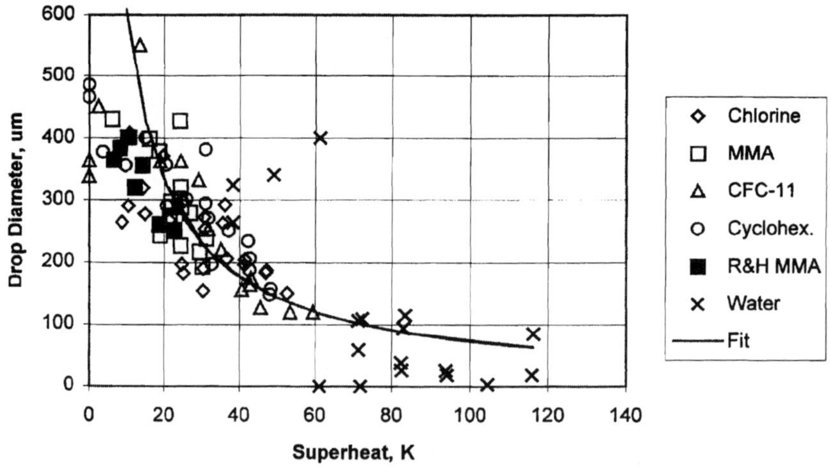

Figure 4.10. Drop size correlated with superheat for flashing liquids.

Superheat, $\Delta T_{sh,}$ is defined as the excess temperature above the saturation temperature, T_s (the boiling point at the applicable atmospheric pressure):

$$\Delta T_{sh} = (T - T_s) \tag{4.31}$$

4.1.6. Rainout

The flammable mass in an aerosol cloud is decreased by rainout of liquid droplets. The effect of rainout is to decrease the source strength of the original vapor cloud and to prolong the event by developing a pool of liquid that continues to evaporate unless ignited to form a pool fire.

A number of rule of thumb correlations have been proposed for providing estimates of the rainout fraction as summarized in Table 4.3. These are based on the flash fraction, x, given by Eq. (4.14) which differs slightly depending on whether the expansion occurs along a path of constant entropy (giving x_S) or constant enthalpy (giving x_H). Discharge through an orifice or nozzle is usually considered isentropic and flow along a pipe is considered isenthalpic, but for the very approximate formulas in Table 4.3 either value of flash fraction is adequate. Here, spray fraction means the mass fraction of aerosol that is suspended and evaporates without raining out.

The rule of thumb formulas are plotted in Figure 4.11. The formulas of Dow, Kletz, Mudan, and Schmidli are similar. The formula of Prugh has a large discontinuity.

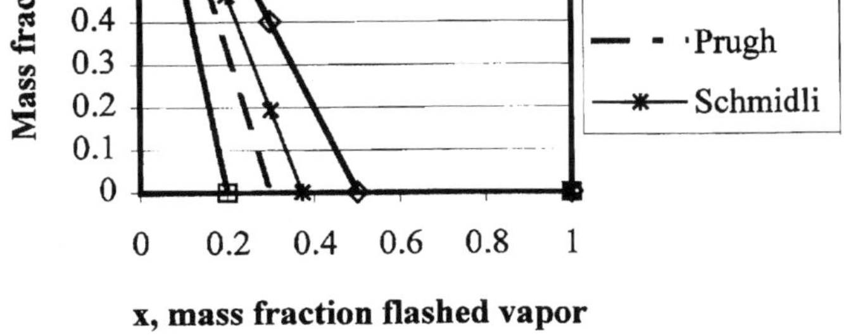

Figure 4.11. Comparison of rules of thumb for rainout.

TABLE 4.3
Some Proposed Rule-of-Thumb Correlations for Predicting Rainout Fraction, η_R

Reference	Calculation
Continuous	
Kletz (1977)	$\eta_R = 1-2x$ if $x < 0.5$ $\eta_R = 0$ if $x > 0.5$ (spray fraction = flash fraction)
Dow (1994)	$\eta_R = 1 - 5x$ if $x < 0.20$ $\eta_R = 0$ if $x > 0.20$
Instant Releases	
Davenport (1977)	$\eta_R = 0$ if $T_0 > T_b$ and $T_b < 21$ C
Prugh (1989)	$\eta_R = 1 - \frac{\min\{x,[(1-x)/2]}{1-x}$
Mudan (1988)	$\eta_R = \begin{cases} 1 & \text{if } x<0.15 \\ 0 & \text{if } x>0.30 \\ \text{linear interpolation} & \\ \text{between} & \text{if } 0.15< x < 0.3 \end{cases}$
Opschoor (1988)	$\eta_R = 0$ if $x > 0.3$
Cox (1990) Same as Kletz (1977) but for instantaneous case	$\eta_R = 1 - 2x$ if $x < 0.5$ $\eta_R = 0$ if $x > 0.5$
Schmidli (1993)	$\eta_R = 1 - 2.69x$ if $x \le 0.37$ for full vessel $\eta_R = 0$ if $x > 0.37$ for full vessel $\eta_R = 1 - 2.18x$ if $x \le 0.46$ for full vessel $\eta_R = 0$ if $x > 0.46$ for full vessel

More physically based models of rainout are treated in the CCPS Concept Series book *RELEASE: A Model with Data to Predict Aerosol Rainout in Accidental Releases* (1998):

- DeVaull and King's correlation
- The RELEASE model
- Rigorous drop evaporation and trajectory models

The DeVaull and King correlation is given by Eqs. (4.32)–(4.35) below. It makes use of the adiabatic saturation temperature, T_{as}. Unfortunately, the

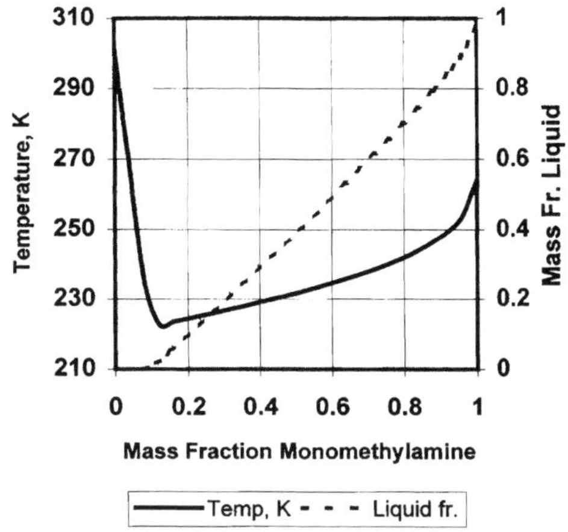

Figure 4.12. Example adiabatic mixing curve showing adiabatic saturation temperature for monomethyl amine.

adiabatic saturation temperature is not widely tabulated, but a simple program can readily generate an adiabatic mixing curve.

This is done by assuming the liquid and vapor phase are at equilibrium, and solving the two conservation equations for mass and energy and the two-phase density given by Eq. (4.12). For a given instantaneous mass of hydrocarbon, m_C, vary the mixing ratio of air, $m_{air}/(m_{air} + m_C)$ over a range of values and calculate the enthalpy of the mixture. As long as liquid exists, find the isentropic vapor flash fraction, x with Eq. (4.14).

An example adiabatic mixing curve is shown in Figure 4.12 for monomethylamine. The minimum temperature is reached as the last drop evaporates, and is the adiabatic saturation temperature. Values for T_{as} are shown in Table

TABLE 4.4
Values Used in Correlation for Rainout

Material	T_s (K)	T_{amb} (K)	T_{as} (K)	$\frac{T_{amb} - T_{as}}{T_{amb}}$	Scaling Factor η_R*
Chlorine	239.1	304.4	199.8	0.344	0.200
Monomethylamine	266.8	301.3	221.8	0.264	0.385
CFC-11	297.0	288.7	241.4	0.164	0.618
Cyclohexane	353.9	308.4	273.1	0.115	0.733
Water	373.2	298.0	288.1	0.057	0.868

4.4 for a few materials, along with the normal boiling point or saturation temperature, T_s. For the materials in Table 4.4, T_{as} is 40 to 85 K below T_s.

The DeVaull and King correlation defines volatile and nonvolatile materials by the criterion that a material is volatile when

$$\frac{T_{amb} - T_{as}}{T_{amb}} \geq 0.14 \tag{4.32}$$

For nonvolatile materials the correlation is

$$\eta_R = 1 - \frac{C_{PL}(T_0 - T_{as})}{h_{fg}} = 1 - x - \frac{C_{PL}(T_s - T_{as})}{h_{fg}} \tag{4.33}$$

Written in the second form, the correlation contains the term $1 - x$, similar in form to the rules of thumb in Table 4.3. For volatile materials the correlation is

$$\frac{\eta_R}{\eta_R^*} = \begin{cases} 1 - \left[\dfrac{x_H}{0.145}\right]^{1.8} & \text{for } x < 0.145 \\ 0 & \text{for } x \geq 0.145 \end{cases} \tag{4.34}$$

The scaling factor η_R^* is defined as

$$\eta_R^* = 1 - 2.33\left(\frac{T_{amb} - T_{as}}{T_{amb}}\right) \tag{4.35}$$

The DeVaull and King correlation is adjusted to the CCPS rainout test data (CCPS, 1998), and may not generalize well for extrapolations beyond the experimental conditions used. In particular, the CCPS test materials were released at an elevation of 1.2 m. However, this correlation does improve upon the rules of thumb correlations as shown in Figures 4.13 and 4.14. Figure 4.13 shows rainout data for cyclohexane (CCPS, 1998) which is classified as nonvolatile at 300 K by the criterion as shown in Table 4.4. The Kletz rule of thumb is plotted as the dashed line in Figure 4.13. Figure 4.14 shows rainout data for water which is similarly classified as nonvolatile. The scaling factor in Eqs. (4.34) and (4.35) is responsible for offsetting the DeVaull and King correlation so that in Figure 4.13 it does not predict 100% rainout at zero superheat as do the rules of thumb in Table 4.3. For nonvolatile materials, the rules of thumb perform adequately, since the offset at zero superheat is less important, as illustrated in Figure 4.14.

The RELEASE model is described in *Measurement and Modeling of Accidental Aerosol Releases* (CCPS, 1998). Rigorous droplet trajectory and evaporation models are described by Kukkonen and Vesala (1991), Melhem (1993), Woodward (1995), Woodward and Papadourakis (1993, 1995), and Woodward,

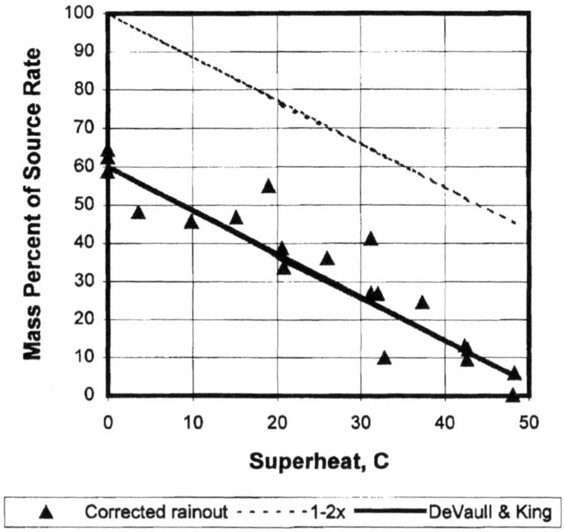

Figure 4.13. Cyclohexane rainout data compared with DeVaull and King correlation for rainout.

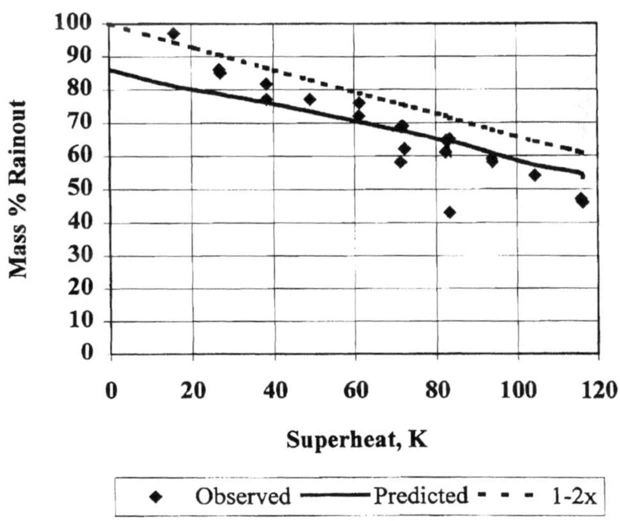

Figure 4.14. Comparison observed rainout data for water with DeVaull and King correlation for rainout.

Cook, and Papadourakis (1995). These models integrate the ordinary differential equations describing droplet momentum, heat and mass transfer to/from the surface of a droplet, and droplet position (trajectories). They obtain droplet temperatures which may decrease below the vapor temperature as the droplets cool by evaporation.

4.1.7. Pool Spread and Evaporation on Land

Materials that rain out also evaporate. Evaporation from a pool of flammable material returns flammable mass to the original cloud. Calculating pool evaporation rates is best done with a computer program. There are too many variables, making an adequate hand calculation difficult, and the evaporation and spread rate are inherently time-varying. Models available in the public domain are listed in Chapter 6. Some models of pool spread and evaporation are compared in the CCPS book *Guidelines for Use of Vapor Cloud Dispersion Models* (1996, pp. 38–45). A discussion of some simple methods is provided here.

The assumption commonly made in modeling rainout is that it occurs at a single point. This is the point at which a representative drop in a monodisperse drop-size distribution would reach the ground. It is also possible to model drop size distributions, and, similarly find the rainout point for each representative drop size. This provides distributed rainout, but since distributed rainout complicates the problem of pool spread and evaporation rate calculation, it is not used in practice.

CONCENTRATIONS ABOVE EVAPORATING POOL

A pool evaporation model must be able to predict not only evaporation rate but also the concentration above the pool. If this concentration is below the LFL, then only a pool fire hazard remains after the original vapor cloud drifts past the pool. A good approximation to the molar concentration near the surface of a pool is found from the vapor pressure, P_v, and the ambient pressure, P_a, by

$$y = P_v/P_a \tag{4.36}$$

One way to calculate concentrations at the surface of the pool with a computer program is to use the same dispersion modeling over an evaporating pool as is used with the vapor cloud. The pool is divided into segments of width Δx along the wind direction. Along each segment, use the evaporation rate model to calculate the rate of flammables entering the cloud. Similarly, use the dispersion model to find the air entrainment rate as a mass flux, in kg of air entrained/(m^2s). Typically, the air entrainment flux and the evaporation flux tend to remain nearly constant, so the concentration of flammable vapors remains nearly constant across the pool.

SIMPLIFIED POOL SPREAD PREDICTIONS ON LAND

A spreading rate equation by Opschoor (1979) is derived by simply equating potential energy to kinetic energy. It predicts spread rate data rather well without having an adjustable constant and without reference to pool viscosity. The radial spread rate, dr/dt, is expressed in terms of the pool height, h, the gravitational constant, g, and an empirical minimum threshold height, h_{min}. The h_{min} value differs with specific models.

$$\frac{dr}{dt} = [2g(h - h_{min})]^{1/2} \tag{4.37}$$

Some simple formulas are summarized for pool spread and evaporation on land by Cox et al. (1990, pp. 46–47). For a continuous release of V (m^3/s), the volumetric flow rate minus the volumetric evaporation rate, the pool area, A_{pool}, is a function of time, t, in seconds up to the duration of the release, t_d, by approximately

$$A_{pool} = \frac{\pi}{4}\left(\frac{2048Vt^3}{81}\right)^{1/2} \quad \text{for } t < t_d \tag{4.38}$$

The pool spreads either until it reaches a confinement, or a minimum average pool thickness, h_{min}, which is typically based on the substrate (i.e., concrete, gravel):

$$A_{pool} = \frac{Vt}{h} \tag{4.39}$$

SIMPLIFIED POOL EVAPORATION RATE FORMULAS

For a volatile liquid, the average evaporation rate flux, G_{evap} (in kg/m^2 s) from a pool on land, at least before the pool temperature cools significantly by evaporation is approximated by Cox et al. (1990):

$$G_{evap} = \lambda u_w \rho_v \frac{P_v}{P_a}\left(\frac{2}{u_w^2 D_p}\right)^{n_c} \tag{4.40}$$

where u_w is the wind speed at a reference height (m/s), ρ_v is the vapor density (kg/m^3), P_v is the vapor pressure, P_a is the ambient pressure, D_p is the pool diameter (m) and λ and n_c are constants dependant on the atmospheric stability class tabulated below. The power law coefficient is defined in terms of the Sutton stability class index, n (an earlier alternative classification to Pasquill–Gifford stability class, see Appendix A), defined by

$$n_c = \frac{n}{2 + n}$$

Pasquill Stability Class	Sutton Index, n	Exponent n_c	Constant, λ
A	0.17	0.078	1.0 E–3
B	0.20	0.091	1.2 E–3
C	0.25	0.111	1.5 E–3
D	0.30	0.130	1.7 E–3
E	0.35	0.149	1.8 E–3
F	0.44	0.180	1.8 E–3

For a cryogenic liquid spilled on land, a simple formula for the evaporation rate flux is (Cox et al., 1990)

$$G_{\text{evap}} = \frac{2Xk}{h_{\text{fg}}}\left(\frac{1}{\pi\alpha}\right)^{1/2}(T_{\text{g}} - T_{\text{L}})t^{1/2} \tag{4.41}$$

where k is the thermal conductivity of the substrate (W/m K), α is the thermal diffusivity of the substrate (m^2/s), h_{fg} is the heat of vaporization (J/kg), T_g is the ground temperature (K), T_L is the liquid temperature (normal boiling point usually), and X is a surface correction factor:

X = 1 for nonporous ground

X = 3 for porous ground because of a greater area for heat transfer

An example of porous ground would be gravel, loose sand, and grassy ground.

COMPUTER PROGRAMS FOR CALCULATING POOL EVAPORATION WITH VARIABLE POOL TEMPERATURE

A computer program calculating pool spread and evaporation can make use of the general heat balance to a pool. This balance determines whether heat loss by evaporation causes the pool temperature to drop, and if so, at what rate it decreases. The net heat rate to a pool, Q_{net}, is the sum of heat inputs by conduction, Q_{cond}, convection, Q_{conv}, radiation, Q_{rad}, reaction with moisture (or for a pool on water, the heat of solution of soluble materials), Q_{sol}, and the enthalpy in the material entering the pool, Q_{leak}, minus the heat of evaporation, Q_{evap}, each in watts or BTU/hr:

$$Q_{\text{net}} = Q_{\text{cond}} + Q_{\text{conv}} + Q_{\text{rad}} + Q_{\text{sol}} + Q_{\text{leak}} - Q_{\text{evap}} \tag{4.42}$$

Each term is developed from standard engineering relationships as summarized in *Measurement and Modeling of Accidental Aerosol Releases* (CCPS, 1998). If the pool temperature is at the boiling point, then the heat loss by evaporation must equal the sum of heat inputs, or

$$Q_{\text{evap}} = Q_{\text{cond}} + Q_{\text{conv}} + Q_{\text{rad}} + Q_{\text{sol}} + Q_{\text{leak}} \tag{4.43}$$

If the heat inputs are less than the heat loss by evaporation, the difference between the two, or the net heat rate, is negative. In this case, the rate of change of pool temperature is calculated from the net rate of heat into the pool, the mass in the pool, m_{pool}, and the heat capacity of the pool material, C_{PL}:

$$\frac{dT_{\text{pool}}}{\text{dt}} = \frac{Q_{\text{net}}}{m_{\text{pool}} C_{\text{PL}} (T_{\text{pool}})} \tag{4.44}$$

Integrating this equation gives the pool temperature. If $Q_{\text{net}} > 0$ the pool will increase in temperature with the constraint that it cannot exceed the boiling point of the liquid.

The pool diameter typically increases until either it reaches a constraint such as a dike or natural obstacle or it reaches a maximum value where the mass loss by evaporation equals the mass input from the spill. Evaporation rate increases at first, as the pool area increases. The evaporation flux remains nearly constant as the pool expands over warm ground, and then drops as the ground and the pool cool. This is illustrated in Figure 4.15 for an instantaneous spill of 10 tons of carbon disulfide on a concrete surface.

The time-variable evaporation rate and pool diameter curves are approximated with discrete elements for dispersion analysis as shown in Figure 4.15.

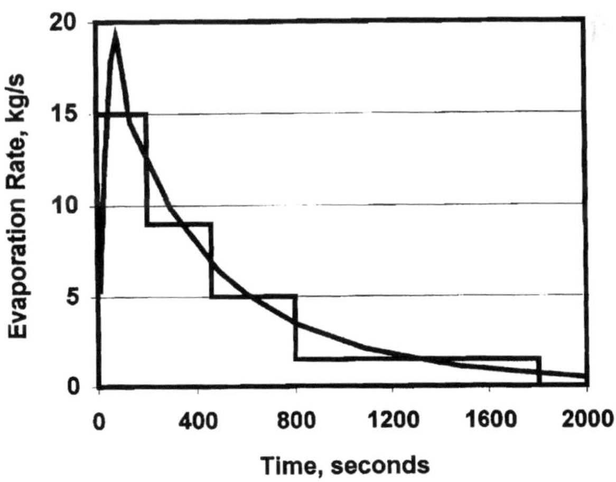

Figure 4.15. Example of pool spread and evaporation from point source of instantaneous spill of 10 tons of carbon disulfide on concrete at 25 C.

One basis for this approximation is to obtain equal evaporated mass in each segment (area under the evaporation rate curve).

Multicomponent pool evaporation is discussed in Appendix F.

4.2. Dispersion Factors

Dispersion can increase or decrease flammable mass in a vapor cloud. It is essential in calculating the flammable mass in a vapor cloud to evaluate how much dilution has occurred, and, if possible, to know the concentration profiles.

Dispersion, or equivalently air entrainment, according to the theory of atmospheric turbulence reviewed by Panofsky and Dutton (1984), is a process of mixing in which turbulent eddies of large wavelength decay into eddies of ever smaller wavelengths.

> *Larger whorls have smaller whorls that feed upon and bite 'em*
> *Smaller whorls have lesser whorls, and so on ad infinitum.*

Dispersion is enhanced by the same factors that promote the acceleration of flame speeds in a fire or deflagration. This section focuses on six important factors that influence dispersion of a flammable vapor cloud

- jet mixing,
- meteorology,
- surface roughness and terrain,
- averaging time,
- impingement,
- wake effects.

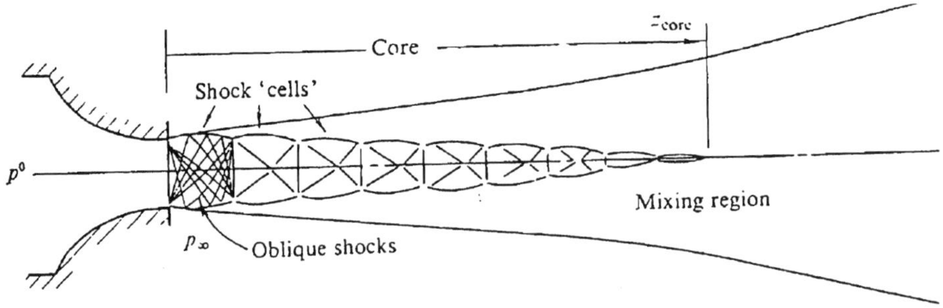

Figure 4.16. Sketch of shock waves developing in underexpanded choked flow jet ($0.48 < P/P_0 < 0.26$) [Reproduced by permission of Cambridge University Press.]

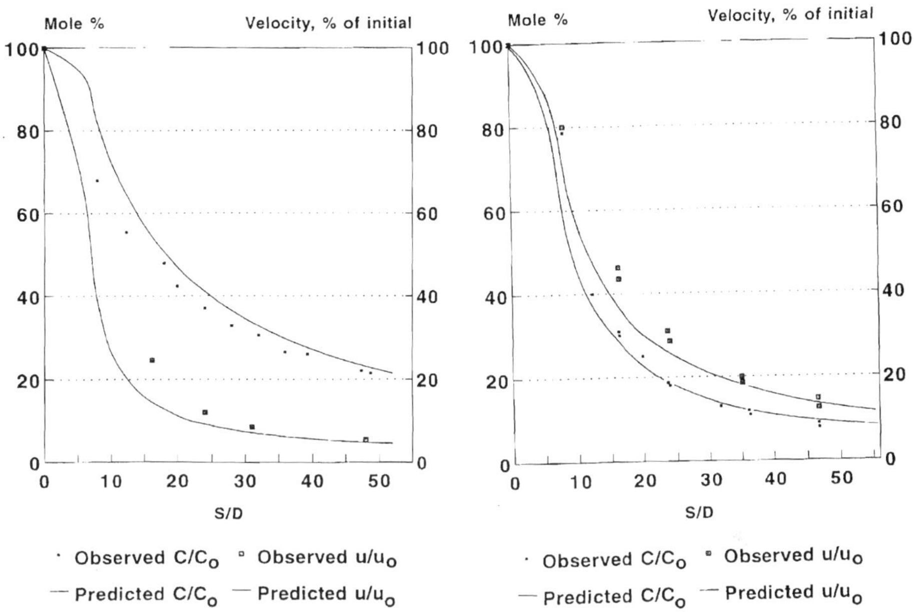

Figure 4.17. Jet mixing data for helium (left) and carbon dioxide (right) discharged in air, described by single-parameter jet model.

4.2.1. Jet Mixing

The largest source of turbulence for mixing a vapor cloud is provided by the pressure energy in the source vessel or pipe converted to kinetic energy upon discharge. With choked flow, expansion takes place to supersonic velocities beyond the choke point and shock waves develop. These shock waves are in a central cone near the source, as illustrated in Figure 4.16. Mixing occurs in the outer shell of the jet. The outer shell is not directly affected by the shock waves, but the momentum transfer in the outer shell rapidly dilutes the plume.

As the plume entrains ambient air mass it also entrains the ambient air's momentum. Thus, the jet velocity and concentration decay at a nearly proportional rate. Since mass and momentum transfer occur by the same mechanism in jet mixing, a successful jet model will need only a single adjustable parameter to provide an adequate fit of both concentration and velocity decay. This is illustrated in Figure 4.17 (Woodward and Cook, 1995) with the data of Keagy and Weller (1949) for a jet into stagnant air. Centerline velocity and concentration decay proportionately and rapidly. The momentum (centerline velocity) decays faster for the low molecular weight helium than for the higher molecular weight

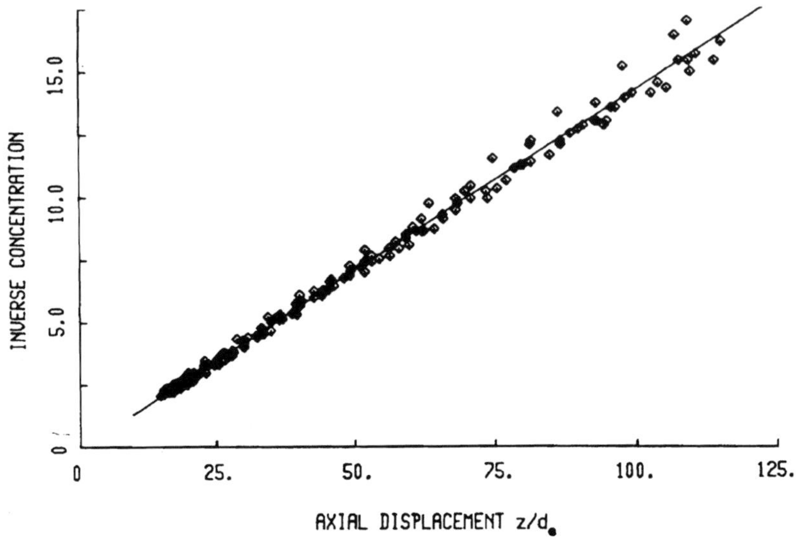

Figure 4.18. Concentration profiles for a 95% methane jet as a function of axial displacement. (Reproduced by permission of Gordon and Breach Science Publishers.)

carbon dioxide. Concentration decays less rapidly with helium which has a lower momentum to force air entrainment.

Concentration profiles from jet mixing are nearly linear when plotted as the reciprocal of mass fraction or mole fraction, as illustrated in Figure 4.18. Figure 4.18 from Birch et al. (1987, citing Birch et al., 1984) plots data for a jet of 95% methane discharged at 20 m/s. The data are plotted as reciprocal mole fraction, y_c^{-1}. The open and closed points refer to different initial pressure.

Rapid dilution is evident in Figures 4.17 and 4.18. A methane jet at the relatively mild discharge rate of 20 m/s reaches twice the LFL of 10%, or the reciprocal mole fraction of 20 in only about 67 discharge diameters, according to Figure 4.18.

A simple formula for the centerline concentration, c_{cl}, ratioed to the source concentration, c_0, for a jet mixing in a stagnant medium (no cross wind) is given by Cox et al. (1990) based on Ricou and Spalding (1962) as

$$\frac{c_{\mathrm{cl}}}{c_0} = 5\left(\frac{\rho_{\mathrm{a}}}{\rho_0}\right)^{0.5} \frac{D_{\mathrm{h}}}{x} \tag{4.45}$$

Here, D_{h} is the source diameter, x is the downstream distance, ρ_0 is the source density, and ρ_{a} is the ambient density. This formula is not valid very near the source.

The bending of jet plumes in a cross wind is described by a second adjustable parameter, as is discussed elsewhere (Rajaratnam, 1976).

4.2.2. Meteorology

The most important meteorological variables that affect dispersion of flammable mass in a vapor cloud are wind speed and atmospheric stability. Precipitation (rain or snow) is also important for removing soluble components such as ammonia, although not normally considered in vapor cloud models. Wind both moves a cloud and generates turbulence that mixes a vapor cloud.

ATMOSPHERIC STABILITY

Atmospheric stability describes the tendency for vertical mixing in the atmosphere and hence for turbulence generation by natural forces. The theoretical basis for atmospheric stability is summarized by Hanna et al. (1982). The atmospheric stability classification scheme of Pasquill and Gifford is in general use. This scheme assigns letters from A through G for increasingly stable atmospheres.

Unstable atmospheres of Pasquill–Gifford Stability A through C typically occur in the daytime hours with bright sunshine. As the ground is heated, convective updrafts generate large-scale turbulence which decays to small-scale turbulence producing enhanced dispersion. Stable atmospheres, Classes E and F, on the other hand, typically occur at night with low wind speed. In fact, Class F occurs only at night. Stabililty Class G applies only over water at night. The ground cools by radiation, the ceiling of the mixing layer lowers, and turbulence is suppressed by the temperature gradient in the atmosphere. Stability Class D represents neutral stability, and occurs most frequently. Overcast conditions with light to strong winds give rise to Stability D.

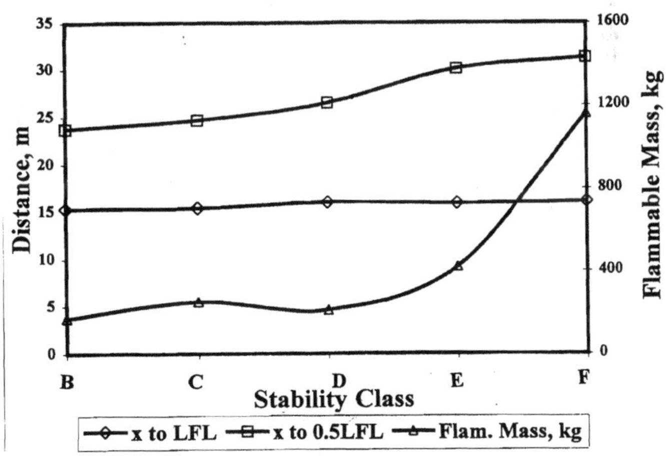

Figure 4.19. Effect of atmospheric stability class on flammable cloud of ethylene oxide release at 25 C, 3.5 m/s wind speed.

Flammable clouds usually behave as dense clouds and spread of their own density in the near field. Later, at concentrations below the LFL, diffusion dominates and in the far field the dispersion becomes passive and more dependent on atmospheric stability. Consequently, flammable clouds from heavy vapors are relatively insensitive to atmospheric stability class as illustrated in Figure 4.19. This plots the flammable mass and cloud length to the LFL and ½ LFL concentrations for a 1000 kg release of ethylene oxide. The distance to the ½ LFL is only slightly sensitive to stability class, and the distance to the LFL is essentially independent of stability class. The flammable mass, though, is higher with stable atmospheres (E and F stability). The importance of stability class is far different for toxic clouds, since their levels of concern extend to much lower concentrations.

Atmospheric stability class must be inferred from several measurable values, and is an inexact science. Several classification algorithms are summarized in Appendix A based on reviews by Sedefian and Bennett (1980), and Gifford (1976). The simplest classification algorithm by (Gifford, 1976) correlates stability class with the readily observed variables of wind speed and cloud cover. However, this algorithm, in Table A.1 in Appendix A, was found to produce inconsistent and varied choices with dispersion tests (McQuaid, 1981). Other methods listed in Appendix A provide better consistency but require additional instrumentation such as a meteorological tower to measure vertical wind speed and temperature profiles. One method, adopted by the US Nuclear Regulatory Commission (NRC), requires measurement and calculation of the standard deviations in the horizontal wind direction, σ_θ, using a bidirectional anemometer.

WIND PROFILES AND WIND SHEAR EFFECTS

Since wind speed increases with height, the effect is to shear a vapor cloud as illustrated in Figure 4.20. The top of the cloud is moved along ahead of the bottom. There is always a residual amount of flammable material at the ground

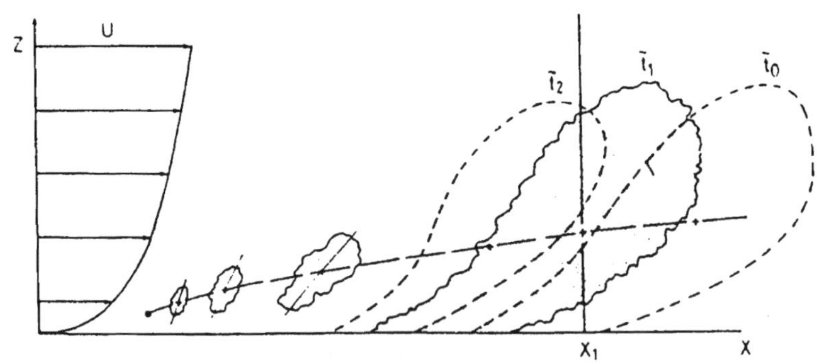

Figure 4.20. Effect of wind shear on a series of puffs (Wilson, 1991, after Dumas et al., 1977). (Reproduced by permission of Academy of Science, Paris.)

level, since boundary layer theory states that the cloud velocity is zero at solid boundaries. The residual ground level concentrations eventually drop below the LFL after the release stops and the cloud passes, but the low wind speeds near the ground provide lower dispersion rates as well as slower advection than wind speeds at higher elevations. Wind speeds cited in meteorological data are conventionally measured at an elevation of 10 m. The pertinent wind speed for flammable cloud dispersion is usually closer to 2 to 4 m.

To correct wind speeds measured at 10 m, $u_w(10)$ or u_{10}, to those at lower elevations, $u_w(z)$, meteorologists describe vertical wind profiles using a logarithmic profile. The logarithmic wind speed profile is given by

$$u_w(z) = \frac{u_*}{k}\left[\ln\left(\frac{z}{z_0}\right) - \Psi\left(\frac{z}{L}\right)\right] \qquad (4.46)$$

Here $k = 0.40$ is the von Karman constant. The term is a function of atmospheric stability class using L, the Monin–Obukov length, as described in Appendix B. For neutral, D stability, Ψ is zero, and Eq. (4.46) is readily solved once two parameter values are known, the roughness length, z_0, and the friction velocity, u_*. The roughness length is discussed in Section 4.2.3. The friction velocity is correlated with atmospheric stability class, or it can be adjusted to fit vertical wind speed profile data.

The logarithmic wind profile has an unfortunate characteristic; it is undefined for heights below the roughness length. For this reason, and for ease of use, dispersion modelers prefer to use a power law profile given by

$$u_w(z) = u_w(10)\left(\frac{z}{10}\right)^p \qquad (4.47)$$

The power law coefficient, p, can be adjusted so that the power law wind profiles closely match the logarithmic wind profiles. Typical values for p range from 0.2 for A stability to 0.5 for F stability. The effect of this range on adjusting wind speeds from 10 m to 1 m elevation is illustrated in Table 4.5.

TABLE 4.5
Example Wind Speed Adjustments for Elevation Wind Speed in Meters per Second

Atmosphere Stability	Elevation	10 m	1 m
A	Power law coefficient $p = 0.2$	10	6.31
F	Power law coefficient $p = 0.5$	10	3.16

For pool evaporation, the wind speed should be evaluated near the pool surface.

4.2.3. Surface Roughness and Terrain

Surface roughness length (height), z_0, is a parameter in the vertical wind speed profiles described in the previous section. Typical similarity dispersion models (in which the concentration profile form is assumed) use surface roughness length as a parameter to affect entrainment rates. Higher values of z_0 also produce higher turbulence and consequently higher dispersion which usually widens the cloud and decreases the flammable mass in a cloud.

TABLE 4.6
Surface Roughness Length Associated with Terrain Descriptions (Pasquill, 1974)

z_0 (m)	Typical Terrain	
1×10^{-4}	Calm open sea or snow-covered flat or rolling ground	Large expanse of water or desert.
1×10^{-3}	Off-sea wind in coastal areas	
2×10^{-3}	Natural snow over farmland	
3×10^{-3}	Frenchmen's Flats, NV test site	
5×10^{-3}	Cut grass (≈30cm)	Fairly level grassy plains
1×10^{-2}	Few trees, winter	
2.5×10^{-2}	Uncut grass, isolated trees	Airport runways
5×10^{-2}	Few trees, summer	Farmland
8×10^{-2}	Many hedges	
3×10^{-2}	**EPA rural cases**	
2×10^{-1}	Many trees, hedges, few buildings	
4×10^{-1}	Outskirts of town	Fairly level wooded country
5×10^{-1}	Centers of small towns	
1.0	Centers of large towns, cities. **EPA urban cases.** **Processing plants** (Peterson, 1990)	Forests
1.5–3.0	Centers of cities with very tall buildings	Very hilly and mountainous

Rigorously, z_0 is found to fit wind speed data over two or more heights. In practice, look up tables, such as Table 4.6, are usually used to select z_0 on the basis of a list of terrain features. For example, dispersion in a wooded area would be expected to be larger than over grassy farmland. In general, $z_0 \approx h/10$, where h is the height of the roughness elements, such as ground cover (Hanna et al., 1982, p. 6).

Surface roughness length applies with obstacles that are small relative to the height of the cloud. Obstacles as large or larger than the height of the plume deflect the plume direction. The effect of such obstacles is only beginning to be modeled, as described in Section 4.2.6. The limitations of the roughness length parametrization were addressed in a wind tunnel study sponsored by the American Petroleum Institute (API). Using scale models of refineries Peterson (1990, 1997])and Peterson et al. (1988) concluded that scaled roughness lengths equivalent to z_0 values of around 1 m are applicable to refinery structures.

Auer (1978) found significant differences in the average meteorological conditions between the suburban and industrial (refinery) areas around St. Louis, Missouri. The average temperature over the industrial area is about 1 C higher and the average humidity is around 1.5 g/kg lower compared with surrounding residential terrain. He draws contours with these differences 3 to 4 km from the boundaries of the industrial areas. Assuming that similar differences apply to the average surface roughness length between suburban and industrial areas, this implies we should use z_0 = 1 m in a zone 3–4 km wider than the industrial area. Beyond that, use roughness lengths typical of the surrounding terrain.

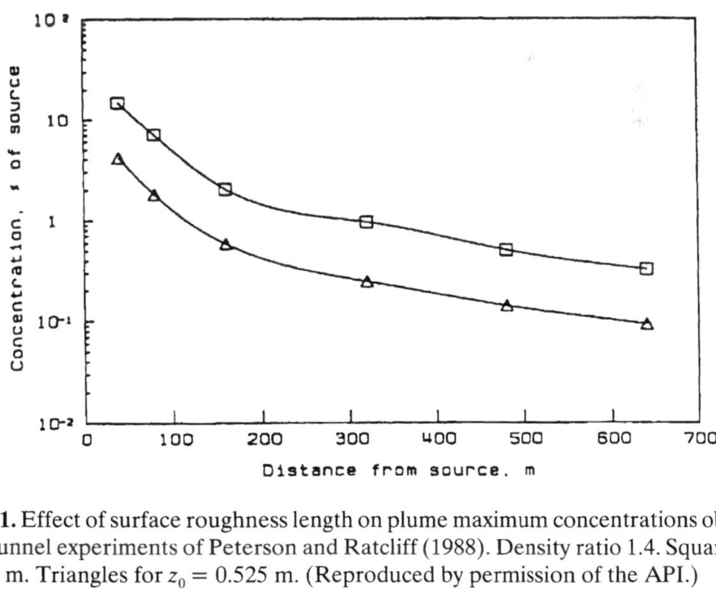

Figure 4.21. Effect of surface roughness length on plume maximum concentrations observed in the wind tunnel experiments of Peterson and Ratcliff (1988). Density ratio 1.4. Squares are for z_0 = 0.015 m. Triangles for z_0 = 0.525 m. (Reproduced by permission of the API.)

Increased surface roughness has a substantial influence on dispersion as illustrated in Figure 4.21. In Figure 4.21, the distance to about 2% concentration (a typical LFL value) is about doubled in going from high roughness to low. A larger flammable mass would be expected over a low roughness length area. The wind tunnel data of Peterson and Ratcliff (1988) also discussed by Roberts et al., (1990) are plotted for medium density gas. They found neutrally buoyant gas concentrations are reduced by a factor of about 3. For heavier gas of density ratio 1.4 concentrations are reduced by a factor of about 4. With a gas of density ratio 3.9 concentration reductions are by a factor of about 5.

The surface roughness length applicable to processing plant structures can be estimated from the dimensions and spacings of structures on the site. The correlation provided by Lettau (1969) is

$$z_0 = \frac{0.5h^* s}{A} \tag{4.48}$$

where h^* is the average obstacle height, A is the horizontal area over which z_0 is to be estimated (m^2), s is the total silhouette area of all obstacles in the area A, measured in a vertical crosswind plane, and 0.5 is the average drag coefficient for the obstacles. When $s/A = 1/5$, Lettau's correlation reduces to the Hanna et al. rule of thumb, $z_0 \approx h/10$. Peterson (1997) evaluated Lettau's formula with wind tunnel studies using refinery models and found it is remarkably accurate.

TERRAIN EFFECTS

The flow of heavy gas clouds follows the terrain much like water runoff from rain. Ravines channel heavy gas flow until the height of the gas cloud exceeds the height of the ravine walls. This has been observed for a spill near drainage channels in The Netherlands. The heavy gas remained confined by the channel walls for miles. Another example was a large-scale natural carbon dioxide release from a volcano at Lake Nyos in Cameroon in August 1986. The gas filled the valley to a certain height. Those living above this height survived, while some 1700 people below the gas level perished (World Map, 1995).

Unfortunately, convenient dispersion models have not yet been developed which describe such important terrain effects as channels and slopes. This is an area for future research as discussed in Chapter 8.

A class of models called shallow-layer models has been developed, as mentioned in Chapter 6, which treat dispersion in some simple terrain features. One such simple terrain is a mild up slope or down slope. The model of Tickle (1996) and of Weber et al. (1993) uses for an instantaneous release a wedge-shaped cloud found from wind tunnel studies of Schatzmann et al. (1990) and of Flacher (1990). These new models show the expected increased dilution down a slope and decreased dispersion flowing up a slope.

4.2.4. Averaging Time

The reported measurement of concentration by a sensor is typically the average over the measurement period, or averaging time. This averages the fluctuations in concentration over this period and also accounts for the instrument response time. There is an averaging time inherently associated with every model for vapor cloud concentrations. For flammable mass calculations, this averaging time should be quite short, 1 to 10 s, consistent with the lag time for ignition by a hot surface. For toxic clouds, longer averaging time is appropriate. With longer averaging time, the reported concentrations are corrected (decreased) from the short averaging time values.

For flammable clouds, averaging time corrections should normally not be needed. The user should insist that his model developer should remove all averaging time corrections for flammable clouds. Usually the only correction needed is for concentrations predicted in the passive (Gaussian) regime. This is basically all a user needs to know. The remainder of this subsection is to cover the subject completely.

The averaging time correction is in two parts: adjustments for cloud duration and cloud meander. A duration adjustment is made for toxic clouds if the *duration* of the passing cloud is less than the *exposure* time used in obtaining toxic dosage effects data. A *cloud meander* adjustment is made in all cases for toxic releases.

A linear *duration adjustment* will report concentrations multiplied by the ratio t_{dur} / t_{basis} to correct from an actual source duration of t_{dur} to an exposure time basis, t_{basis}. Clearly, this correction decreases reported concentrations, which is appropriate for toxic effects, but not for flammable effects. Duration adjusted concentrations should not be used to find flammable mass.

Cloud meander occurs because of wind direction changes during the course of a release. A sensor or observer at a fixed point near the ground could experience periods of discernable concentrations, followed by periods of essentially zero concentration or intermittency. The average concentration observed is decreased by the intermittency.

Some heavy gas models use fast sensor data for the near-field, heavy-gas dispersion, and passive (Gaussian) dispersion for the far field. Passive dispersion data are taken with averaging times typically of 10 to 30 min. Thus, if a flammable cloud is predicted to extend into the passive regime which calculate, say 30 minute average concentrations, these predicted concentrations should be increased to give consistently short time averaged concentrations.

If the model reports predictions based on a meander-adjusted averaging time, τ_2, that is larger than $\tau_1 = 10$ s, these predictions should be corrected to the shorter averaging time. The traditional adjustment uses the 0.2 power law relationship relating concentrations, c, and cloud width (expressed as a standard deviation σ_y, with the averaging times

$$\frac{c(\tau_1)}{c(\tau_2)} = \frac{\sigma_y(\tau_2)}{\sigma_y(\tau_1)} = \left[\frac{\tau_2}{\tau_1}\right]^{0.2} \tag{4.49}$$

Wilson, in a CCPS Concept Series book, *Concentration Fluctuations and Averaging Time in Vapor Clouds* (1996) evaluated the experimental basis for the 0.2 power law and found that a better expression was needed which takes into account other important variables. Namely, he found a dependence of averaging time on source size (diameter), and on the normal spectrum of atmospheric turbulence.

4.2.5. Impingement and Cratering

Discharges often impinge upon nearby surfaces such as the ground, a wall, a tank, or piping. Discharges from a broken buried pipe often form a crater. In impinging on a surface or the walls of a crater, the momentum of the jet and corresponding entrainment rates are considerably reduced. The flammable mass can be considerably larger for an impinging jet than for a free jet. Impingement also increases rainout. Small droplets or solid particles are able to follow the streamlines illustrated in Figure 4.22 and remain suspended, but large drops and particles impinge upon surfaces and rain out.

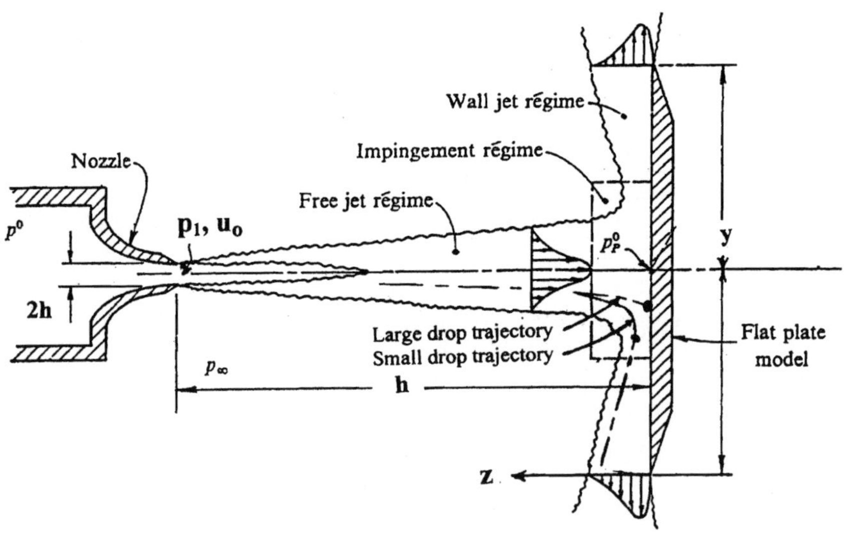

Figure 4.22. Flow profiles and droplet trajectories for two-phase jet impinging on a flat surface. Impingement angle shown as 90° (Reproduced by permission of Cambridge University Press.)

Commonly used dispersion models do not account for such factors as impingement and wake effects. Some simple geometries have been studied, both experimentally and using computational fluid dynamics models. One useful case is a jet directed downward obliquely or normal to a flat plate, as illustrated in Figure 4.22. Three regimes are defined: (1) the free jet regime where the jet properties are free of any interaction effects due to impingement; (2) the impingement regime where the strong interaction of the jet with the impingement surface produces a change in flow direction; (3) the wall jet regime consisting of essentially radial flow along the impingement surface. The impingement and wall jet regimes where found by Donaldson and Snedeker [1971] to occur when the flat plate is located within 50 times the jet nozzle radius ($h/b < 50$).

Improved models of flammable jets can be developed using a comprehensive review of experimental data and computational fluid dynamics (CFD) modeling predictions of Loonie and Walsh (1984). Their results for a free turbulent jet and jets impacting a flat plate normally apply with a wide range of flow variables.

Flow from bursting underground pipes that form craters is discussed by Wilson (1981). He suggests reducing the momentum of the discharging jet by a term with a drag coefficient times the area of the crater.

4.2.6. Obstacle Effects

Present similarity-type dispersion models do not account for the presence of obstacles in the path of a vapor cloud. Obstacles can include pipe racks and pumps as well as large columns and tanks. Furthermore, fin-fan coolers, cooling towers, and furnaces are known to generate localized wind patterns in their immediate vicinity. Wind tunnel studies have shown that vapor clouds become wider and shorter upon passing through typical plant structure [Peterson, 1989]. A cloud that is predicted by conventional models to pass through several units may actually pass through far fewer units before diluting to LFL concentrations

Turbulence induced by obstacles increases dilution and accelerates or quenches flame speeds. Large obstacles (buildings) develop a wake that can be approximated as a well-mixed zone. Wind tunnel studies reveal the flow patterns around isolated block-like buildings. The predominant features of these flow patterns are sketched in Figure 4.23 (Hunt et al., 1977; Woo et al., 1977). In a recirculation wake zone downwind of the obstacle, the flow on average is reversed. Turbulence is high in this zone leading to fairly uniform concentration profiles typical of a well-mixed zone. The length of this recirculation zone is about three times the building height, H. No effect is seen after about 10 times H. Hosker and Pendergrass (1987) define buildings as tall when the height exceeds the width, $H > W$, and as squat when $W > H$. They provide a procedure for

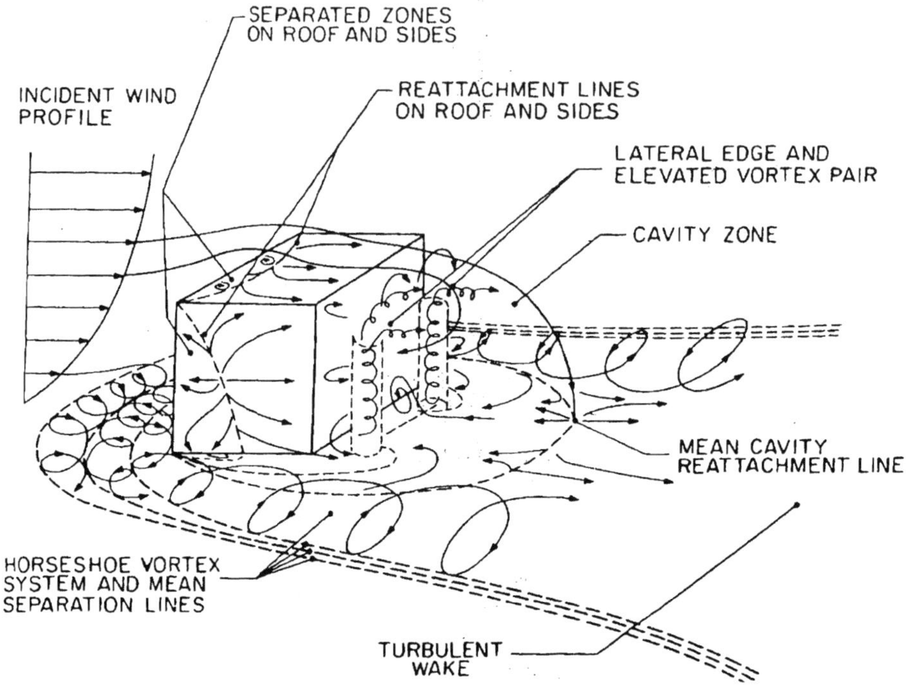

Figure 4.23. Flow patterns around an isolated cube in the turbulent boundary layer (Hunt et al. 1977; Woo et al. 1977).

modifying the standard deviations of plume width, σ_y and plume height, σ_z at distances $\geq 3W$ for tall buildings and $\geq 3H$ for squat buildings.

At the downwind side of the recirculation zone, the wind flow reattaches to the ground, but for some distance downwind the mean velocity remains lower than its upwind value at the same height above the ground. All along this wake turbulence is higher than its upwind values. The stagnation streamline intersects the building above the ground. Below this streamline there is a separated flow region in which a strong vortex is wrapped around the body to form the "horseshoe" vortex. Thus, flammable material near a building can have higher concentrations in the building wake than in the absence of the building. Enhanced turbulence accelerates flames near and far downwind of the building.

Ongoing research is focusing on developing simplified algorithms to describe the overall effects of simplified block-type obstacles on the main properties of a plume: concentration, enthalpy, and momentum. An example of such simplification is provided by Duijm et al. (1995) who assume that the cloud can

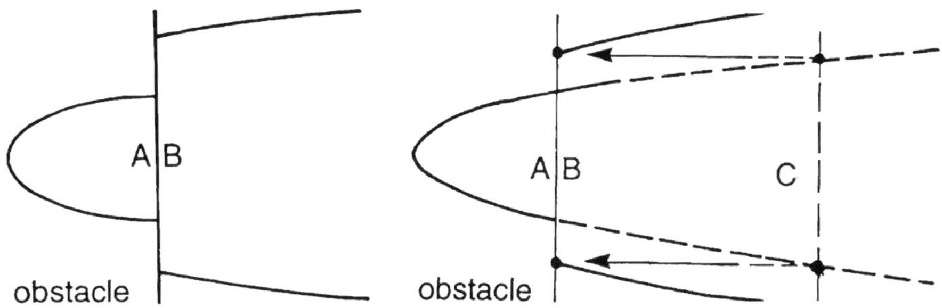

Figure 4.24. (*Left*) Plan view sketch of cloud outline according to the obstacle model of Duijm et al. (1995). (*Right*) Procedure of Duijm et al. (1995) for incorporating the obstacle model in a flat terrain dispersion model. (Reproduced by permission of Elsevier Science Publishers.)

be described by a discontinuity as it passes an obstacle. As illustrated in Figure 4.24A,B, plume properties are modeled as over uniformly flat ground up to point A immediately upwind of the obstacle. The effective height, width, and concentration of the cloud at point A (h_A, w_A, c_A) are considered to change discontinuously around the obstacle to some other set (h_B, w_B, c_B) immediately downstream of the obstacle. After it passes the obstacle further evolution of the cloud continues as it would over uniform flat terrain.

The conditions at point B are found by formulating a set of "dilution criteria" at point A. Then, ignoring the obstacle, suppress all model output and let the model evolve until it reaches point C where the dilution criteria are satisfied. The conditions at point C (h_C, w_C, c_C) are then made the conditions at point B (h_B, w_B, c_B). Effectively the dilution criteria result in substituting a virtual source identical in character to the original source, but displaced by a distance ($x_c - x_B$), as illustrated in the right panel of Figure 4.24.

The dilution criteria make use of an estimated turbulence length scale or eddy size L_e appropriate to the obstacle dimensions (height, width, length H_b, W_b, L_b):

$$L_e = \min(\alpha W_b, \beta H_b)$$

(or similar power law functions of building dimensions)

The dilution criteria are

1. Continuity: For top hat model profiles (well-mixed, no gradients) $h_A = h_c$, $w_A = w_c$, $c_A = c_c$.
2. Aspect ratio of cloud is assumed unchanged in going from point B to point C (obstacle has no effect on this).
3. The cloud height must extend over the obstacle. Also the cloud width must be larger than the turbulence length scale. Alternately, a lateral

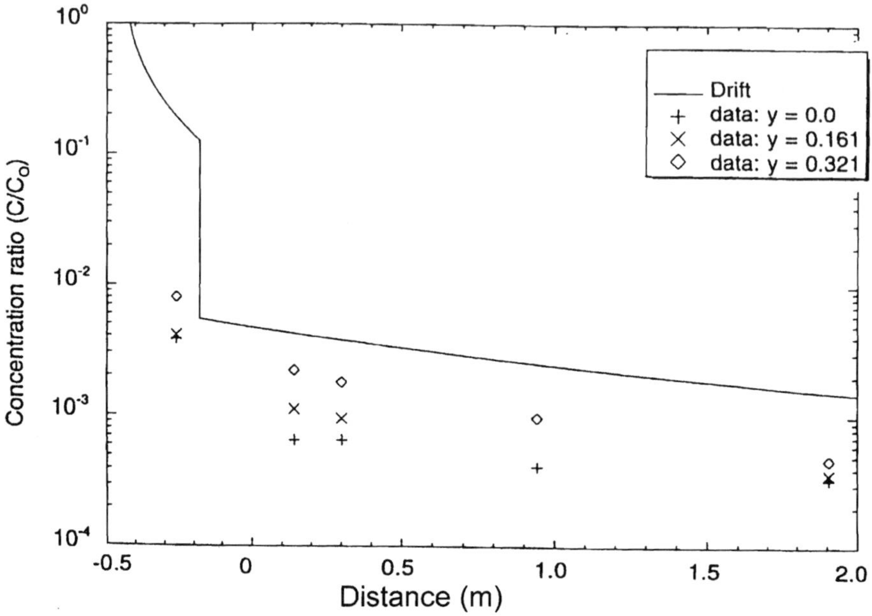

Figure 4.25. Comparison of obstacle model of Duijm et al. (1995) with wind tunnel tracer gas data for a medium-width building $W_b = 3H_b$. (Reproduced by permission of Elsevier Science Publishers.)

condition must be satisfied which depends on the transverse position of the plume and obstacle. That is

$$(h_B \geq H_b \quad \text{or} \quad w_B \geq L_e) \qquad \text{OR} \qquad (\textit{Lateral} \text{ condition})$$

The lateral condition is satisfied when part or all of the cloud extends beyond the building. If only part extends around, then only half of the cloud is enlarged by the obstacle.

An example of model predictions compared with wind tunnel data for tracer gas SF_6 around a scaled model building is shown in Figure 4.25 from Duijm et al. (1995). For a medium-width building the model predictions compared with data are fair, as shown. For a wide building, the model matches data better.

A similar model by Cleaver et al. (1995) increases the cloud width and height proportional to obstacle dimensions, but does not introduce discontinuities in concentration. An example of Cleaver's model predictions is shown in Figure 4.26 compared with a field test that used a semicircular fence 50 m from the release. The fence is seen to produce a very nearly discontinuous disturbance

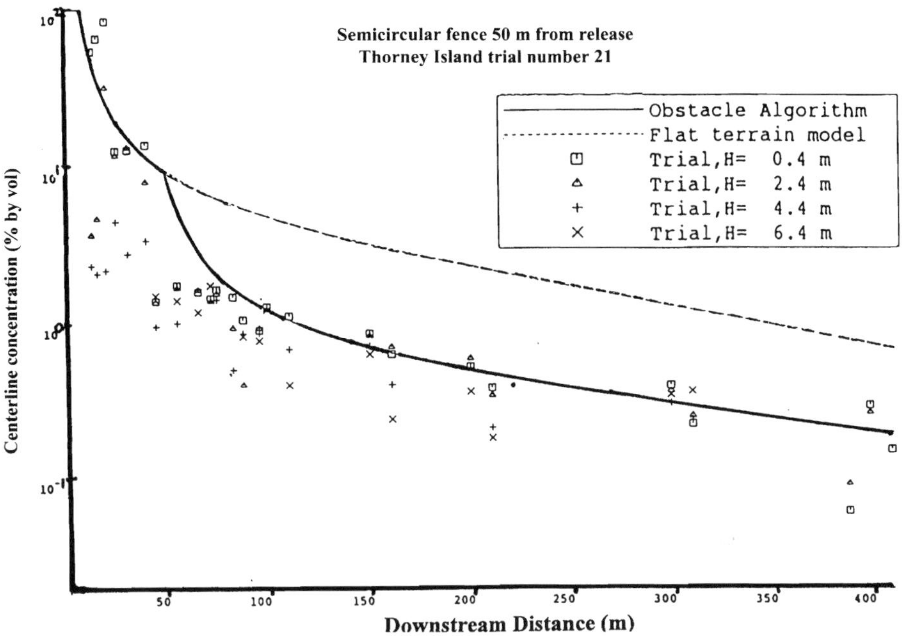

Figure 4.26. Example predictions of obstacle algorithm of Cleaver et al. (1995) applied to instantaneous release of freon–nitrogen dense-gas mixture at ambient temperature. (Reproduced by permission of Elsevier Science Publishers.)

to the concentration profile compared with the profile without a fence. The model predictions also decrease rapidly at the fence, and the fit with data is adequate. The concentration decrease caused by a fence is enough to drop concentrations well below flammable limits if the fence is in the right location and if flow is through the fence rather than over the fence.

Extensive wind tunnel data are provided on the effects of vapor fences and various types of obstacles by Peterson (1995).

4.3. Sources of Ignition

In the chemical industry, ignition sources are the only things we get for free.
—TREVOR KLETZ

The following sources of ignition are recognized by authoritative works (NFPA, 1991):

- Open flames (includes furnaces, smoking, and gas welding or cutting)
- Electrical equipment (including electrical welding or cutting and communications equipment)
- Electrostatic discharge (includes lightning and induction from lightning or charged clouds passing)
- Autoignition and hot surfaces
- Impact or friction (includes thermite reaction, grinding sparks, and rock spalling)
- Pyrophoric materials and runaway reactions

Some sources distinguish between "hard" or high-strength and "soft" or low-strength ignition sources, depending on the ignition energy they generate. According to TNO (Yellow Book, 1997) or Kinsella (1993) only the first two would be considered hard sources. The distinction is blurred, though, for lightning and some runaway reactions can be considered "hard" sources.

Some of these ignition sources are not obvious. Open flames are particularly effective ignition sources since the temperature is high and free radicals generated by combustion are already present. Electrical equipment can generate ignition by either inductive or capacitive discharge or by simply heating by resistance until the temperature exceeds the autoignition temperature. Electrostatic discharge can develop by the charge separation inherent in flowing drops or particles. Lightning develops from this mechanism as drops are circulated in thunderstorm clouds. Hot surfaces serve to heat a portion of a flammable cloud to its

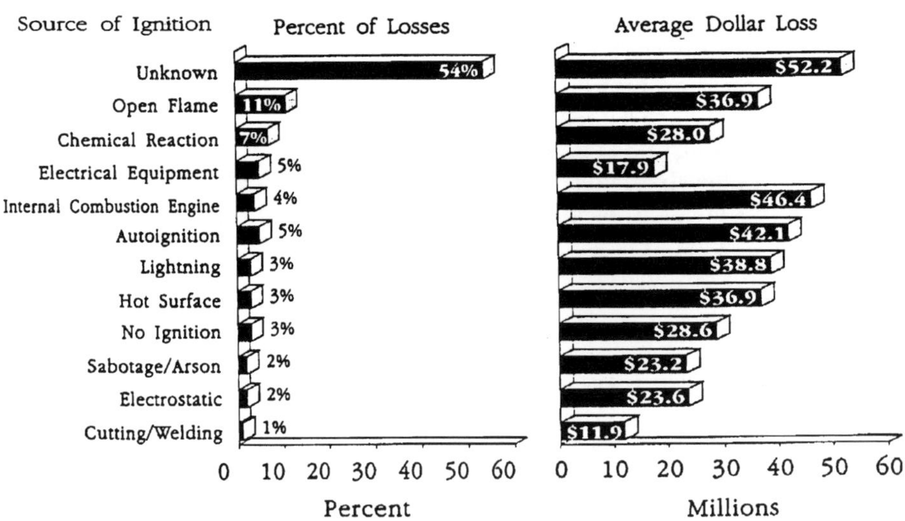

Figure 4.27. Sources of ignition for 170 large fire and explosion events. (Reproduced by permission of J&H Marsh and McLennon.)

autoignition temperature (AIT) and may also be catalytic, causing ignition at a temperature well below the AIT. Impact and friction can generate heat or sparks which are special cases of hot surfaces. Rusty iron tools impacting an aluminum or magnesium metal surface can generate a thermite reaction at the point of contact. Dry metal sulfides are examples of pyrophoric materials which oxidize rapidly upon contact with air. Any such oxidation is a special case of a runaway reaction, of which there are numerous examples. Peroxides and nitrates are particularly energetic ignition sources by reaction.

Figure 4.27 (J&H Marsh and McLennon, 1992) shows the distribution of known ignition sources for 170 major fire and explosion incidents in the petrochemical industry. The leading known source of ignition is open flames, commonly found in furnaces and boilers. The seven incidents in which an internal combustion engine provided the source of ignition produced the highest average loss in this study.

Importantly, in most major incidents the ignition source cannot be identified. Furthermore, not enough is known about ignition sources to fully eliminate them. Thus, for safe design the recommended premise is to assume that sufficiently strong ignition sources are always present.

4.4. Flame Characteristics

Man is born unto trouble, as the sparks fly upward.
—Job 5:7

Calculating flammable mass requires an understanding of a few basic combustion properties. Fire and explosions are basically a rapid reaction between a fuel and an oxidizer in the vapor phase. All flames are gas-phase, characterized as propagated as a wave. The heat of reaction raises the temperature, vaporizes material from the surface of solids or liquids, and generates a feedback mechanism which accelerates and perpetuates the reaction. Since vaporization is an important step for burning liquids, burning rates correlate better with the heat of vaporization than with the vapor pressure of liquids (Eichhorn, 1955; Mudan, 1989).

Flames are known to be reactions in which the fuel is consumed in a chain reaction of free radicals. There are pre-mixed and diffusion flames. In either case, ignition starts in a premixed zone and perpetuates as a wave until it reaches a fuel-rich zone (a diffusion zone). One model of the reaction–diffusion zone of a flame, assuming large activation energies for the chain-branching reaction, consists of a thin fuel consumption layer where radicals are formed and an even thinner radical consumption layer. Energy from the reaction zone is transferred to the preheat zone ahead of the reaction wave. At steady-state, the chain-breaking reaction equals that of the chain-branching reaction (Lewis and von Elbe, 1951; Peters and Smooke, 1985). The rate-limiting step in diffusion flames is air/fuel mixing, so turbulence is important. Turbulence can quench a flame or accelerate it, depending on the fuel concentrations gradients nearby (blowing out a candle removes heat by increasing dilution and by turbulence, fanning smoldering embers increases available heat). Local heat loss by radiation or conduction is important in propagating flames, particularly near the flammable limits. These properties are applied in the design of flame, deflagration, and detonation arresters in which the flame front is forced to pass through a metal mesh of small gap widths.

Ignition can be by two ways. The first is by a small source such as a spark or flame at a point in a flammable mixture. The second is by heating the bulk gas to

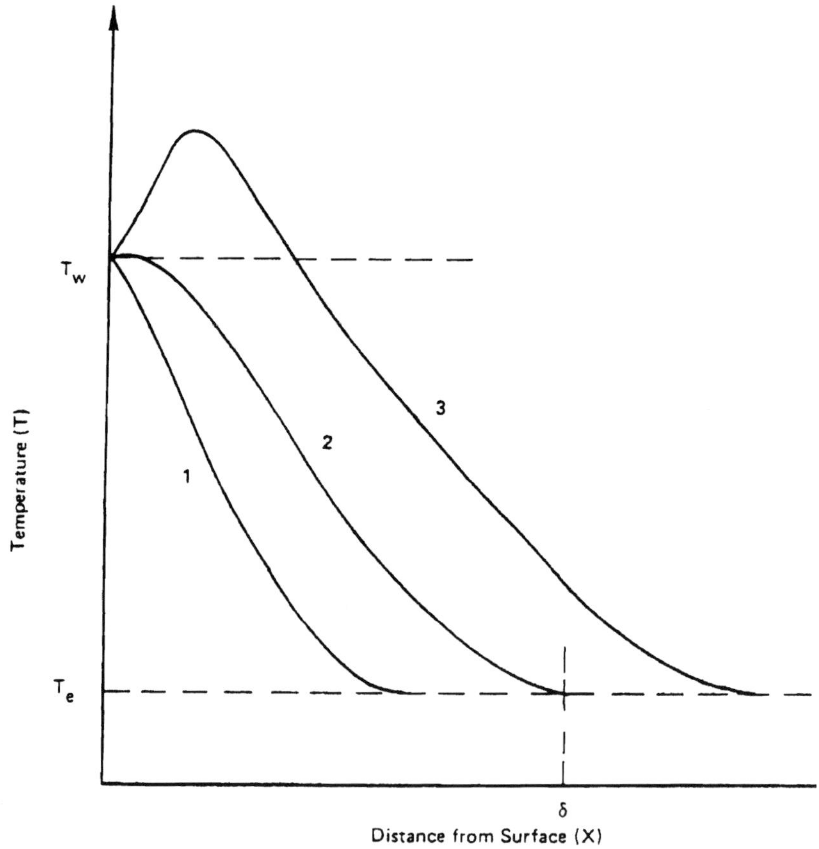

Figure 4.28. Temperature profiles near heated surface surrounded by a flammable mixture. Legend: (1) quenching flames; (2) ignition; (3) after ignition. From Laurendeau (1982). (Reproduced by permission of The Combustion Institute.)

its autoignition temperature. A fire needs to be supported by an ignition source until it is self-sustaining. An example cited by Lees (1996) is that thermite bombs which reach 1650 C for 15–20 s, have been found to be less effective in starting fires than napalm, which reaches about 1100 C for 10–15 min.

Ignition by hot surfaces occurs when the energy loss from the surface, q_{loss}, equals the energy gain by chemical reaction, q_{chem}. That is, the temperature profile at ignition is perpendicular to a vertical hot surface as illustrated in Figure 4.28. The lower line in Figure 4.28 represents flame-quenching conditions, and the upper line represents postignition conditions. The heat loss term is given by standard heat transfer theory using the temperature difference between the wall, T_{w}, and the gas, T_{g}, the thermal conductivity of the gas, k, a characteristic length of the surface, L, and the heat transfer coefficient, h, given by the Nusselt Number, Nu ($=hL/k$), which is found using standard correlations:

$$q_{\text{loss}} = \frac{kNu}{L}(T_{\text{w}} - T_{\text{g}}) \tag{4.50}$$

The chemical reaction rate is given by a standard kinetics expression with the mole fraction of the fuel, y_{F}, mole fraction of the oxidizer, y_{O}, the mixture density, ρ_{cld}, a kinetics frequency factor, A, an activation energy, E, and the partial orders of the reaction, m_{f}, and m_{O}, making up the overall reaction order $n = m_{\text{f}} + m_{\text{O}}$:

$$r_{\text{f}} = -(y_{\text{f}})^{m_{\text{f}}}(y_{\text{O}})^{m_{\text{O}}}\rho_{\text{cld}}Ae^{-E/RT} \tag{4.51}$$

The heat gain from chemical reaction at the wall is the product of thermal conductivity and temperature gradient at the wall:

$$q_{\text{chem}} = k\left(\frac{dT}{dx}\right)_{\text{wall}} \tag{4.52}$$

and the temperature gradient is shown by Laurendaeau (1982) to be readily found from the heat balance equation using the heat of combustion, H_{c}:

$$k\frac{d^2T}{dx^2} + H_{\text{c}}r_{\text{f}} = 0 \tag{4.53}$$

Ignition energies are lowest and laminar flame speeds are highest near stoichiometric concentrations. This is because the energy density upon combustion is highest when there is just enough oxygen for complete combustion. The stoichiometric concentration is readily found by writing the formula for combustion of a general hypothetical component as follows (with X = fluorine or chlorine):

$$\begin{aligned} &C_jH_kO_LS_mN_nP_pX_q + \beta(O_2 + 3.773N_2) \\ &= jCO_2 + \tfrac{1}{2}(L-q)H_2O + mSO_2 + (\tfrac{1}{2}n + 3.773\beta)N_2 + qHX + \tfrac{1}{4}pP_4O_{10} \end{aligned} \tag{4.54}$$

where 3.773 = 0.7905/0.2095. The stoichiometric concentration in mole fraction is

$$y_{st} = \frac{1}{1 + 4.773\beta} \tag{4.55}$$

$$\beta = j + \tfrac{1}{4}(k - q) + m + \tfrac{5}{4}p - \tfrac{1}{2}L \tag{4.56}$$

The combustion energy per unit volume at stoichiometric concentrations for many of the most common flammable gases is fairly equal at about 3.5 MJ/m^3 of fuel–air mixture. Some quick methods for estimating flammable mass take advantage of this fact. Another consequence of this fact is that the adiabatic flame temperature of most fuels at stoichiometric concentrations in air is about 2370 K.

Another relatively constant property of hydrocarbon fires is the volume expansion ratio upon combustion, r_V, or volume of products per unit volume of reactants. The expansion ratio, is an important property driving the development of an explosion. It ranges between about 6 and 8. If the combustion products are not allowed to expand freely, a pressure increase will take place. For total confinement (constant volume enclosure) the pressure increase is, therefore, a factor of about 8. The expansion ratio depends upon two factors, the density ratio before/after combustion, ρ_0/ρ_b, and the mole ratio of products and reactants r_m:

$$r_v = r_m \frac{\rho_0}{\rho_b} = r_m \frac{T_b}{T_0} \frac{\sum n_{bi} M_{bi}}{\sum n_{ri} M_{ri}} \tag{4.57}$$

The ideal gas law is used for the densities. This factors to the temperature ratio after/before combustion, T_b/T_0 and the molecular weight ratio of combustion products to reactants. Equation (4.64) gives the enthalpy balance for calculating T_b (also termed AFT). The ratio r_m is readily found, for example, from the formula for complete combustion of a hydrocarbon in dry air:

$$C_jH_k + \beta O_2 + 3.773\beta N_2 = jCO_2 + \frac{k}{2}H_2O + 3.773\beta N_2 + \alpha O_2$$

where

$$\alpha = \beta - j - \frac{k}{4} \tag{4.58}$$

The ratio of moles of products to moles of reactants is

$$r_m = \frac{j + k/2 + 3.773\beta + \alpha}{1 + \beta + 3.773\beta} \tag{4.59}$$

At the stoichiometric concentration, $\alpha_s = j + 0.25k$. Since for C_3 and higher saturated homologs, $k = 2j + 2$, and for and unsaturated homologs, $k = 2j$, the numerator is almost equal to the denominator, and r_m has a typical value around 1.05. The equivalence ratio, ϕ, or the degree of approach to stoichiometric, is α/α_s.

4.4.1. Flammable Limits

The lower flammable limit (LFL) is the premixed fuel concentration below which the fuel–air mixture is too lean for self-propagation as a flame. Above the upper flammable limit (UFL) it is too rich. The LFL and UFL are also called the lower explosive limit (LEL) and upper explosive limit (UEL). Between the LFL and the UFL is the flammable range. There is a narrower range in which detonations can occur. Lewis and Von Elbe (1951) tabulate some detonatability limits. Flammable limits are tabulated in Appendix C. See also Kohlen (1992). Bolk (1998) describes the flammable envelope as a function of the Reynolds number.

Flammable limits are not a fundamental property, but rather a characteristic that depends on the measurement protocol. For example, low emission engines operate in a region with the average concentration below the nominal LFL. Catalytic combustion can occur at concentrations below the LFL. Furthermore, for aerosols and dusts, the LFL depends on aerosol droplet size and dust particle size, as discussed in Section 4.5.

The most commonly used test for the determination of flammable limits is the Bureau of Mines apparatus described by Coward and Jones (1952). This consists of a tube 5 cm in diameter and 1.5 m high sealed at the top and open at the bottom. A small ignition source is placed at the bottom. The flammable limits are concentrations at which a flame just travels the full length of the tube.

Appendix C lists flammable properties by a coarse reactivity classification commonly used for explosion modeling. This classification, suggested by Zeeuwen and Wiekema (1978) is based on the laminar flame speed as discussed in Section 5.3.3. Appendix C shows that most organic vapors have a flammable range between about 1 volume or mole % and 15%. Oxygenated materials have a wider range. Some highly reactive materials (including acetylene and ethylene oxide) have a UFLs up to 100%. For paraffins, Jones (1938) noted that the LFL is close to 55% of the stoichiometric concentration (at $\varepsilon_r \cong 0.5$), and similarly the UFL is about twice the stoichiometric concentration (at $\varepsilon_r \cong 2$).

Flammability limits for hydrocarbons are generally widened with increasing pressure as illustrated for methane in Figure 4.29 (Zabetakis, 1965b). The limit values vary linearly with the logarithm of pressure up to 700 bar. However, for ethanol and 1-propanol Gibbon et al. (1994) found the LFL increases linearly with pressure up to 7 bar (100 psig).

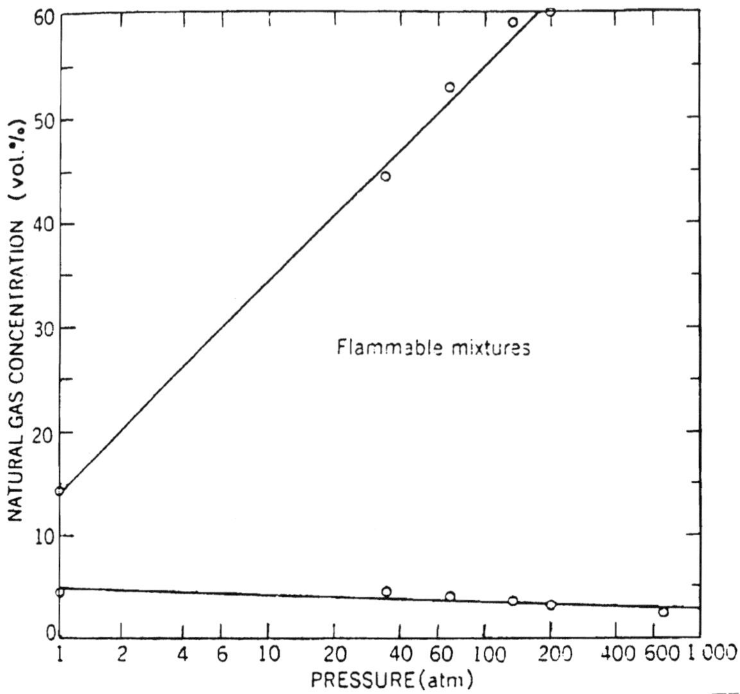

Figure 4.29. Effect of pressure on the LFL and UFL of methane and air at 25 C.

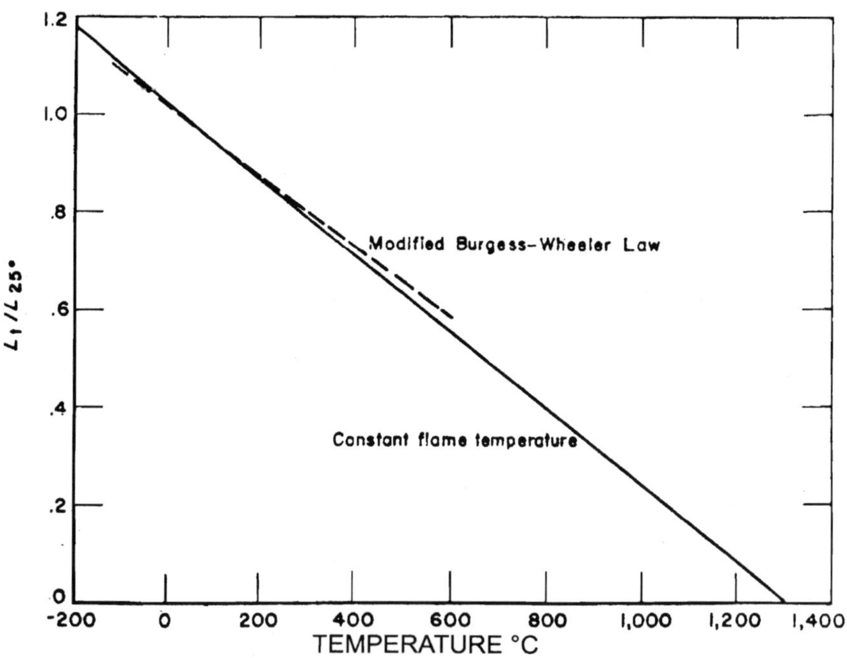

Figure 4.30. Effect of temperature on the LFL illustrating the modified Burgess–Wheeler law. (Reproduced by permission of U. S. Bureau of Mines.)

Flammability limits are also widened by increasing temperature. In fact, the effect of temperature on the LFL and UFL is linear. Figure 4.30 illustrates the linearity of the LFL for a constant flame temperature. The LFL is expressed by the modified Burgess–Wheeler law (Zabetakis, Lambiris, and Scott, 1959) in terms of the heat of combustion, H_c in kJ/kg. A fairly good approximation for hydrocarbons is

$$\text{LFL}(T) = \text{LFL}(25\ \text{C}) - \frac{3142(T-25)}{H_c} \tag{4.60}$$

For example, for *n*-hexane the heat of combustion is 3.852×10^6 kJ/kg. The LFL at 25 C is 1.2 vol. %. At –200 C Eq. (4.60) gives a value 1.15 for LFL(T)/LFL(25 C) and at 400 C this ratio is 0.755. These points fall on a line with a slightly lower slope than shown in Figure 4.30.

The Burgess–Wheeler law applies also for the UFL except for the UFL the coefficient is positive.

With pressures below atmospheric, the LFL and UFL come together at a point and below that pressure the mixture will not propagate flame. Figure 4.31 illustrates this for the highly reactive ethylene oxide that has a very wide flammable range. The upper flammable limit is 100% because ethylene oxide contains highly strained epoxide bonds. It burns at high concentrations and temperatures

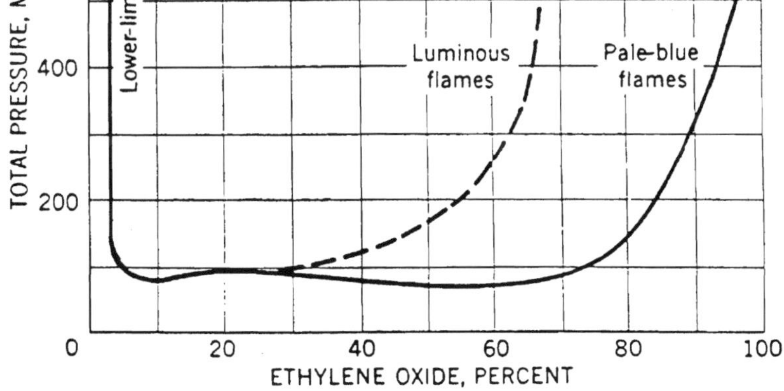

Figure 4.31. Effect of low pressures on the LFL & UFL of ethylene oxide showing range of decomposition flame. (Reproduced by permission of U. S. Bureau of Mines.)

by a decomposition flame. Exothermic decomposition occurs at about 1200 K to produce methane and carbon monoxide. These reaction products can also burn if oxygen is present, increasing the adiabatic flame temperature to above 1600 K. The decomposition flame is nearly colorless and is not extinguished by smothering the flame since it does not need external oxygen. Thus, a pool fire of ethylene oxide may appear to be extinguished, say by a foam blanket in the annular space between the tank and a dike. However, a nearly invisible decomposition flame continues to burn near the hot metal of the tank, and this flame reignites the pool fire when the foam blanket decays. Decomposition flames occur with other materials as well, such as acetylene and propylene oxide.

The LFL and UFL may be predicted for mixtures of flammable components using LeChatelier's rule (LeChatelier, 1891, cited in Crowl and Louvar, 1990) in terms of the vapor phase mole fraction of each flammable component, y_i, and the individual component LFL_i or UFL_i values.

$$\text{LFL} = \frac{\sum_i y_i}{\sum_i (y_i/\text{LFL}_i)} \qquad \text{UFL} = \frac{\sum_i y_i}{\sum_i (y_i/\text{UFL}_i)} \tag{4.61}$$

LeChatelier's rule is empirical. Its limitations are discussed by Coward and Jones (1952). For example, given the following mixture:

Component	Vapor Mole %, y_i	LFL Mole %
Acetone	5.0	2.6
Methanol	12.0	5.9
Ethanol	7.0	3.3
$\sum y_i$	24.0	

$$\text{LFL} = \frac{24.0}{\frac{5.0}{2.6} + \frac{12.0}{5.9} + \frac{7.0}{3.3}} = 3.95\%$$

4.4.2. Flammable Limits with Inerts

LeChatelier's law does not apply for estimating flammable limits when the fuel–air mixture contains noncombustible inerts. Experimental data are available for a number of materials in the form of plots such as Figure 4.32 for propylene. This plot shows the flammable range decreasing with the addition of inerts until above some initial inerts concentration the mixture is not flammable. The main effect of adding inerts is on the UFL which is decreased essentially linearly in proportion to added inerts. The flammable limits curve is brought more to the

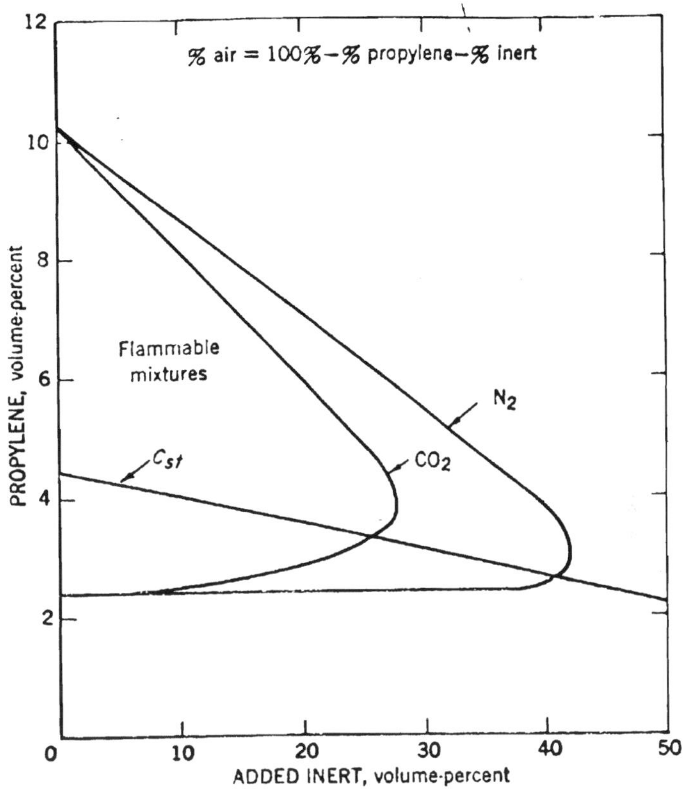

Figure 4.32. Flammable envelope for propylene, air, and inerts nitrogen and carbon dioxide. (Reproduced by permission of the Bureau of Mines.)

left with higher molecular weight inerts (which act as better fire surpressants). The stoichiometric concentration is also linearly decreased by inert concentration. The nose of the flammable limits curve (called the extinction limit) falls very nearly along the stoichiometric concentration line.

When such experimental data are available, simply evaluate whether the mixture composition falls within or without the flammable curve such as Figure 4.32. However, when no such data are available, a method is needed to predict whether the mixture is flammable or not.

Early efforts to correlate the flammable limits for these cases were based on observations by Burgess and Wheeler (1911), LeChatelier (1891) and LeChatelier and Boudonard (1898) that under typical ambient conditions the energy per unit volume of fuel–air mixture of many materials is nearly constant at their limit conditions because it represents the minimum energy needed to propagate a flame. However, in Table 4.7, the heat of combustion per unit volume, H_{cv}, is seen to be just as nearly constant at the stoichiometric concentration as it is at the LFL. (This assertion is made in Section 4.8.4.) H_{cv}, is found as the product of the net heat of combustion (kJ/kg), the molecular weight, M_i, the molar density, $\widetilde{\rho}$, and the mole fraction at a concentration, y_i

$$H_{cv}(y_i) = H_c y_i M_i \widetilde{\rho} \tag{4.62}$$

TABLE 4.7
Typical Physical Properties

Component	LFL, mole %	Stoich., mole %	Mole Wt.	Net Ht. Comb. MJ/kg	Heat of Comb. at Stoich. kJ/m³	Heat of Comb.at LFL. kJ/m³
Methane	5.0	9.48	16.04	27.83	912.3	1730
Methanol	5.9	12.25	32.04	19.8	1530	3176
Ethylene	2.7	6.53	28.05	47.15	1460	3530
Propane	1.7	4.98	44.09	46.16	1414	4143
n-Hexane	1.1	2.16	86.18	44.70	1732	3401

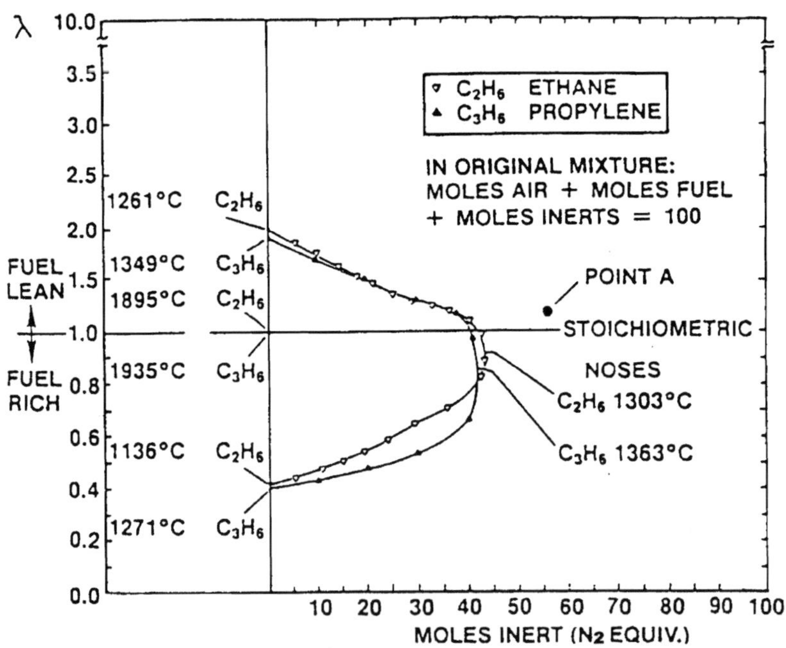

Figure 4.33. Flammable envelope for ethane and propylene showing adiabatic flame temperatures.

A similar assertion is that the adiabatic flame temperature (AFT) calculated from the heat of combustion is also nearly constant at the LFL. Calculating the AFT accounts both for initial temperature and also for the inerts in the original gas mixture. Hansel et al. (1992) expanded on the observation of White (1922), and Edgerton and Powling (1948) that flame temperatures of lower molecular weight paraffins at their LFL and UFL are relatively constant near 1250 C. For example, Figure 4.33 shows how adiabatic flame temperatures drop off on either side of the stoichiometric concentration. Figure 4.33 replots data from Figure 4.32 (and a similar plot for ethylene) as

$$\lambda = \frac{\text{Oxygen mole fraction in fuel-air mixture}}{\text{Stoichiometric oxygen mole fraction}} \tag{4.63}$$

plotted against the mole fraction of inerts expressed as nitrogen (weighted by gas heat capacities). The LFL is the upper line in Figure 4.33, since λ is an inverse relationship to that plotted in Figure 4.32.

As an example of the AFT method consider a composition at 20 C and atmospheric pressure with the following composition:

Component	Mole % Before Comb.	Mole % After Comb.	C_{pvap} (kJ/mole K)
Ethane	0.48	0	
Propylene	1.45	0	
Air	43.60	—	
Oxygen	9.134	0.920	33.73
Nitrogen	48.20 + 34.466 in air	81.876	31.40
Carbon dioxide	4.82	10.033	48.56
Argon	1.45	1.436	20.93
Water	0.0	5.735	38.93
Total	100	100	

Using the average heat capacities over the temperature range 0 to 1100 C (Hansel et al., 1982) with values taken from the JANAF tables (Stull and Prophet, 1971) the AFT is found by an enthalpy balance given by

$$\sum_i w_{ri} H_{ri}(T_{amb}) + H_{ci} = \sum_j w_{pj} C_{pj}(\text{AFT} - T_{amb}) \tag{4.64}$$

Here, w_{ri} is the mass fraction of reactant i before combustion, H_{ri} is the enthalpy of each reactant at ambient temperature, H_{ci} is the heat of combustion, in J/kg, w_{pj} is the mass fraction of products j with average heat capacity C_{pj}, and the reference temperature is taken as ambient, T_{amb}. The AFT is found to be 1065 C.

Since the AFT is <1250 C, the method predicts that the example mixture is not flammable.

This can be confirmed in this example case by finding that the coordinates for this mixture in Figure 4.33, which fall at Point A, are outside the flammable limits. By convention, the nitrogen in the original air is not included in the abscissa. Added nitrogen in the test mixture is averaged on a heat capacity basis with the other inerts to give

$$\text{Mole percent inerts as } N_2 \text{ equivalent} = 48.2 + 4.8\left(\frac{48.5}{31.4}\right) + 1.45\left(\frac{20.9}{31.4}\right) = 56.6$$

This is the abscissa value used for Point A.

4.4.3. Autoignition Temperature for Gases

When a flammable mixture is heated uniformly, there is a temperature at which the oxidation reaction accelerates to ignition, and it is termed the autoignition temperature (AIT). When calculating the conditions in a jet emission, if dilution below the UFL leaves the plume temperature ≥AIT then the plume would probably become a jet fire. This is not entirely certain since there is a measurable and often significant time delay in ignition by hot surfaces. This helps explain delayed ignition events and nonignition.

Autoignition is a function of the hot area, the time of exposure, and the mixture composition. It has been shown that to ignite a methane–air mixture in a cold container, a hot patch of area at least 18 mm^2 is required at 1000–1100 C (Rae, Singh, and Danson, 1964). This is, in effect, an ignition energy requirement. The AIT tests are designed to have a large hot area. The AIT values normally cited are taken in laboratory equipment with clean surfaces. Autoignition temperatures may be reduced by as much as 100–200 C with dusty or rough surfaces. They can be decreased significantly (10–100 C) by testing in larger vessels (Snee, 1988). Rusty iron, nickel, etc. can be catalytic and give lower AIT values.

The delay time or induction time for ignition was shown by Snee and Griffeths (1989) to stretch beyond 22 min in a 0.02-m^3 vessel. This shows that delayed ignition is highly likely for ignition by hot surfaces.

AIT is found by ASTM Method E659-78 which specifies that material in a 500-mL flask is placed in a preheated oven in a darkened room for 10 min, and autoignition is evident by the sudden appearance of a flame inside the flask and an increase in the temperature of the gas mixture. The ignition delay time is also noted. In some cases reactions begin at a lower temperature giving cool-flame ignition with little luminescence and lower temperature increases. Cool-flame ignition is discussed by Lees (1980). The inherent variability in this type of measurement is evident.

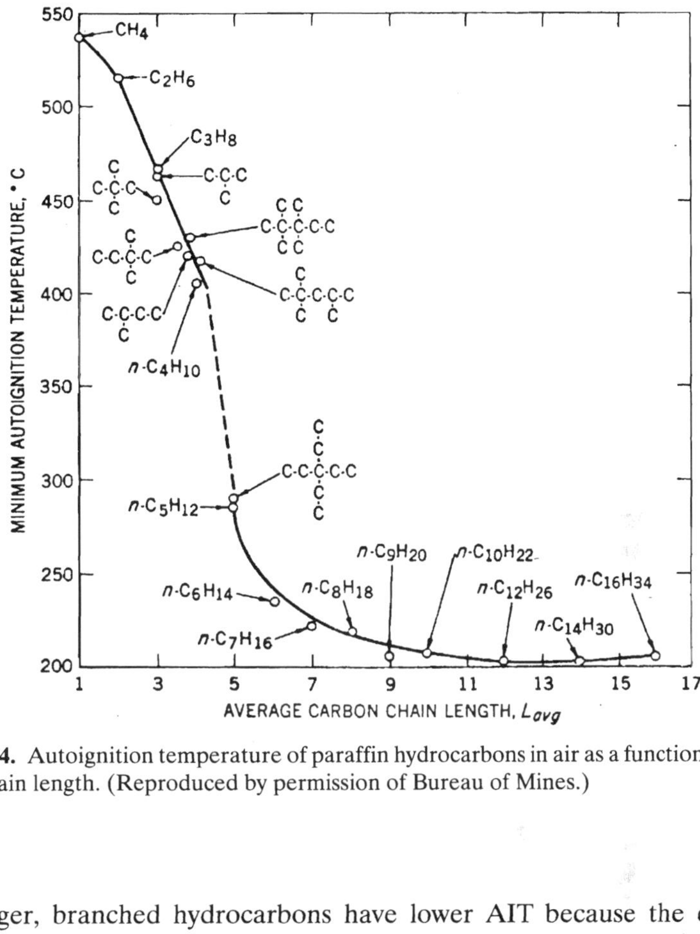

Figure 4.34. Autoignition temperature of paraffin hydrocarbons in air as a function of average carbon chain length. (Reproduced by permission of Bureau of Mines.)

Larger, branched hydrocarbons have lower AIT because the energy to remove a free radical is lower. Figures 4.34 and 4.35 (Zabetakis, 1965) illustrates how AIT decreases with average chain length for paraffins and aromatics with paraffinic chains. The average chain length is defined by the number of chains, ch_i, each containing N_i carbon atoms, and the number of methyl groups, M:

$$L_{\text{ave}} = \frac{2\sum ch_i N_i}{M(M-1)} \tag{4.65}$$

A benzene ring is treated as a methyl group. When the benzene ring contains two side groups, L_{ave} is determined for the side group that gives the largest value, and to this is added 1/2, 1/3, or 1/4 of the average chain length for the second side group (1/2, 1/3, and 1/4 correspond to the ortho-, meta-, and para-positions, respectively).

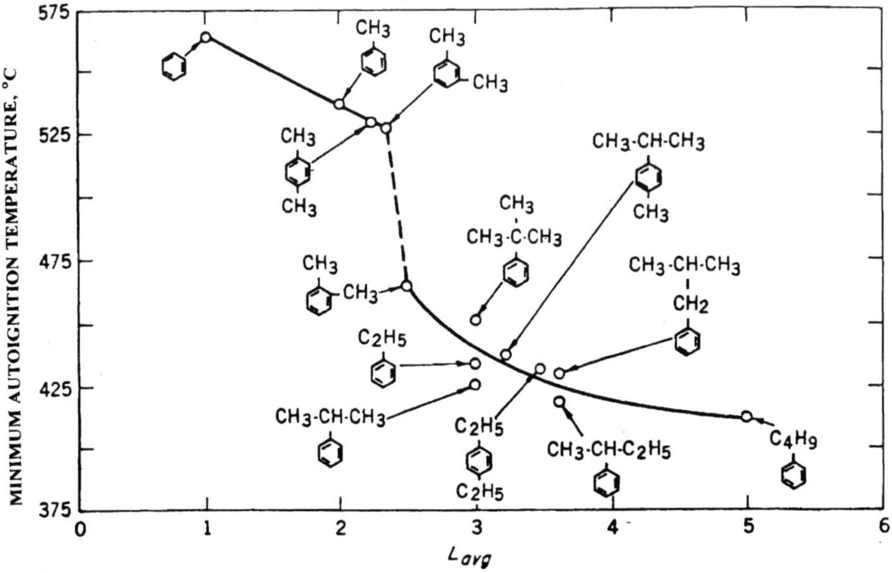

Figure 4.35. Autoignition temperature of aromatic hydrocarbons as a function of the average carbon chain length. (Reproduced by permission of Bureau of Mines.)

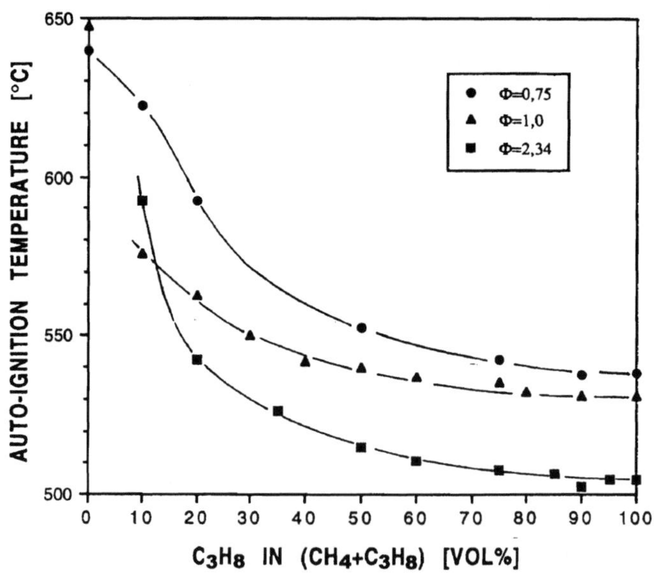

Figure 4.36. Influence of larger molecules in reducing the autoignition temperature of mixtures of hydrocarbons. (Reproduced by permission of Elsevier Science Publishers.)

A small amount of longer or branched hydrocarbon in a mixture lowers the AIT significantly as illustrated in Figure 4.36 (Kong et al., 1995). This is consistent with free radical attack on the weakest link in the carbon chain.

4.4.4. Minimum Ignition Energy for Gases

The flammable mass in a cloud depends strongly on the delayed ignition time. Large hot surfaces can delay ignition measurably, as seen in the previous subsection. Delayed ignition also occurs for ignition by sparks, for different reasons.

When a flammable gas is ignited by a *local* source of ignition, it is not sufficient to raise a volume of the gas above a certain temperature for a certain time. It is also necessary to have a minimum volume of the mixture (the flame kernel) that will give rise to a flame through the rest of the material. This describes ignition by sparks, hot slag from cutting operations, etc. The spark generates a "tunnel" of free radicals in passing through a flammable mixture. The probability of a given set of spark paths resulting in ignition depends on the length of the paths, the energy and speed of the sparks, and on the minimum ignition energy of the fuel.

Electrical sparks are usually used to find the threshold energy for ignition or the minimum ignition energy. Some materials such as hydrogen, acetylene, and carbon disulfide have a very low minimum ignition energy as listed in Appendix C. The minimum ignition energy occurs near stoichiometric concentration such as in Figure 4.37. There is an order of magnitude difference in minimum ignition energy for the energetic hydrogen and the low-reactivity methane.

Flame speeds vs. composition plots have a similar form to Figure 4.37 (see Section 4.4.6). The reason that the center of an explosion is often separate from the edge of the flammable cloud is explained in part by these characteristics of ignition energy and flame speed.

Inherently safe electrical equipment is designed to produce spark energies less than the minimum ignition energy. To put Figure 4.37 and Appendix C values in perspective, the energy in a typical static spark a person experiences in touching a grounded door knob is about 20 mJ. The minimum ignition energies listed in Appendix C are typically <1 mJ. Thus, grounding measures must be carefully designed if flammable atmospheres are expected.

Minimum ignition energies are difficult to measure. There are some practical problems in measuring ignition energy, since narrow spark gaps act as a flame arrester and quench incipient flames. The optimum gap that gives the lowest ignition energy is about twice the maximum quenching distance. When the gap width differs widely from the optimum, up to ten times as much energy may be needed for ignition. With dusts, large differences occur between inductive and capacitive discharge as well.

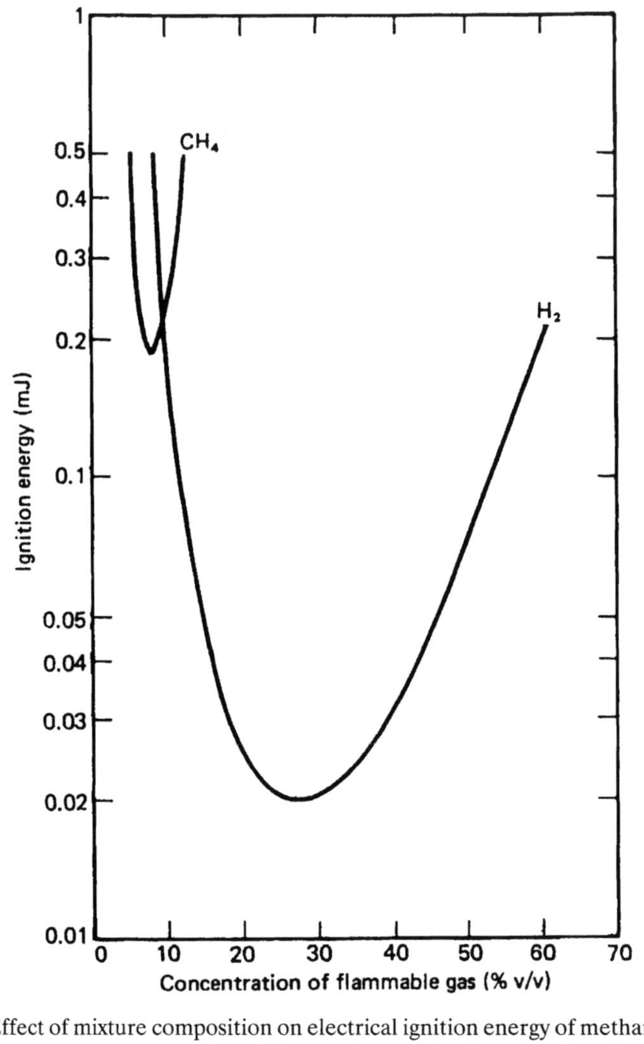

Figure 4.37. Effect of mixture composition on electrical ignition energy of methane and hydrogen in air. (Reproduced by permission of Factory Mutual Engineering Corp.)

4.4.5. Flash Point

Flash point is defined as the lowest temperature at which a momentary flash will develop above a liquid when it is exposed to a pilot flame ignition source. It is essentially the temperature at which the vapor pressure produces a mixture with air at the LFL. Figure 4.38 shows the flash point as the intersection of the LFL and the vapor pressure curve. Flash point is an indicator of the hazardous nature of a particular material for fire or explosion.

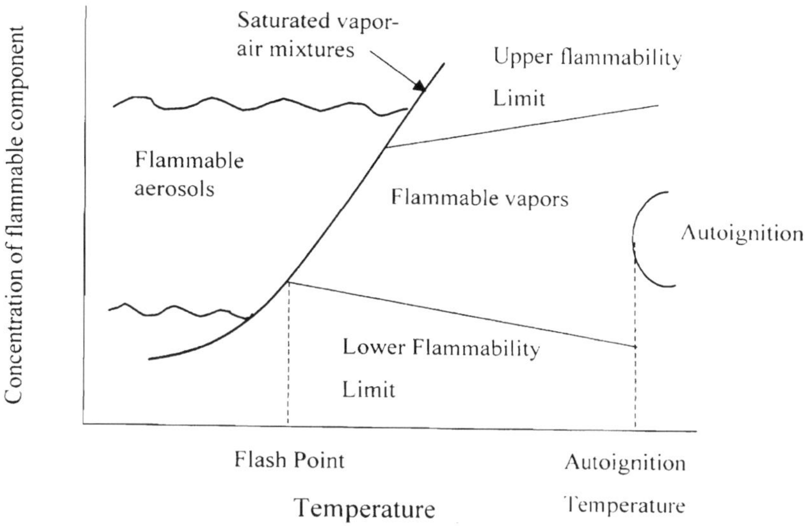

Figure 4.38. Relationships between vapor pressure, flammability limits, flashpoint and autoignition temperature.

There are two categories of flash point, defined by the test method: the open-cup and closed-cup. Closed-cup flash points are a little lower than open cup and are generally regarded as giving a much more accurate indication of the flammability of a liquid. A list of flash point measurement techniques are given in NFPA 30.

Materials become more hazardous when they are heated to temperatures above their flash point. When in closed containers, the vapor composition of a material with a flash point well below ambient temperature is likely to be above the UFL, that is, too rich to burn (e.g., gasoline, flash point –43 C held at 20 C). With liquids which have a flash point above ambient, the vapor composition above the liquid is likely to be too lean. If the flash point is just below ambient temperature then the vapor–air mixture is likely to be within the flammable range and therefore potentially hazardous.

Appendix D provides a correlation for flash point by chemical type. Appendix C tabulates flash point values.

4.4.6. Laminar Burning Velocity and Turbulent Flame Speed

Laminar burning velocity is a characteristic related to the reactivity of materials for explosion and fire modeling. It is measured with a premixed air and fuel establishing a stationary flame in a Bunsen burner. The flame propagation is

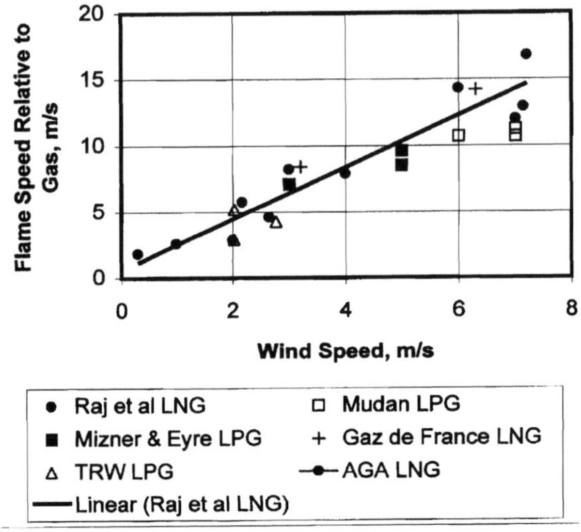

Figure 4.39. Turbulent flame speeds as a function of wind speed for LNG and LPG data.

controlled by diffusion rates of oxygen to the fuel and of combustion products from the flame. The volumetric rate of fuel gas consumption divided by the area of the flame cone is called the laminar burning velocity or fundamental burning velocity, S_u (Lees, 1997, p. 16/22). Correlations for other flame properties are often expressed in terms of S_u. Values of S_u range typically from 0.4 to 1.0 m/s.

Practical combustion problems involve turbulence, and the turbulent flame speed, S_t. Turbulent flame speeds range from low multiples of the laminar flame speed to detonation velocities (500–1000 m/s) or more. Values of S_u are listed in Appendix C and in standard texts, for example, Lewis and Von Elbe (1961), Gaydon and Wolfhard (1970).

Turbulent flame speeds are a function of the expansion ratio and of environmental conditions which generate turbulence. For example, Figure 4.39 illustrates how flame speeds for LNG and LPG clouds increase linearly with wind speed (Mudan, 1989; Rew et al., 1995). In the absence of obstacles, the turbulent flame speeds are nearly the same for LNG and LPG, roughly the wind speed plus a small multiple of their fundamental burning velocity, S_u. Since S_u and expansion ratios are nearly the same for methane and propane, similar turbulent flame speeds would be expected.

4.5. Aerosol Flammability

Mists of small drops of combustible liquids well below their flash points can be fully as explosive as vapor–air mixtures (Eichhorn, 1955; Green and Dressel,

1989). However, high overpressures are usually not experienced with the high molecular weight aerosol explosions (Krawitz et al., 1964).

As discussed in Section 4.1.5, aerosols can form by hot nonvolatile vapors forming a condensation fog, or by pressurized subcooled liquids such as diesel fuel or heat transfer fluids shattering to form small droplets from a leak in a pumped system. Aerosols can also form from releases of a volatile liquid stored above its boiling point that flashes upon release.

The LFL and UFL for aerosols can be expressed as mole fraction, y_i, mass fraction, w_i, or concentration, c_i (kg/m^3). The interrelationship between these values are the general two-phase relationships as follows:

$$c_i = w_i \rho_{\text{cld}} \text{ (concentration = mass fraction × two-phase density)} \tag{4.66}$$

$$y_i = w_i M_T / M_i \text{ (mole fraction = mass fraction × molecular wt. ratio)} \tag{4.67}$$

$$\frac{1}{M_T} = \sum_i \frac{w_i}{M_i} = \sum \frac{w_{vi}}{M_i} + \sum \frac{w_{Li}}{M_i} \tag{4.68}$$

or

$$M_T = \sum_i y_i M_i \tag{4.69}$$

The two-phase cloud density is found with Eq. (4.12) in which the mass fraction of vapor is

$$x = \sum w_{vi}$$

If y_{vi} is the mole fraction of component i in the vapor phase (the vapor-phase LFL), y_{Li} is mole fraction of component i in the liquid phase (usually 1.0), and x' is the mole fraction of vapor, then

$$y_i = x' y_{vi} + (1 - x') y_{Li} \tag{4.70}$$

The aerosol LFL, y_i, will be a slightly different value (expressed as moles of i/total moles of liquid and vapor) than the vapor-phase LFL (expressed as moles of i/moles of vapor). The two-phase LFL reduces to the vapor-phase LFL when $x' = 1$. Importantly, a fuel-rich vapor that is above the UFL can become flammable through condensation of the fuel. The vapor-phase concentration can decrease below the UFL, creating a serious explosion hazard.

Aerosol mists have been shown to have a minimum ignition energy for drop sizes in the range 20 to 30 m as illustrated in Figure 4.40 (from Singh et al., 1985) for tetralin. This is a range in which the drops do not readily settle. Two curves are plotted, for 20% and 60% of the tests that ignited. Figure 4.41 (from Danis et al., 1988) gives similar results for n-heptane as a function of drop size and equivalence ratio.

In about the same drop size range, Britton (1991) reported that aerosols have a maximum laminar burning velocity. The burning velocity decreases with

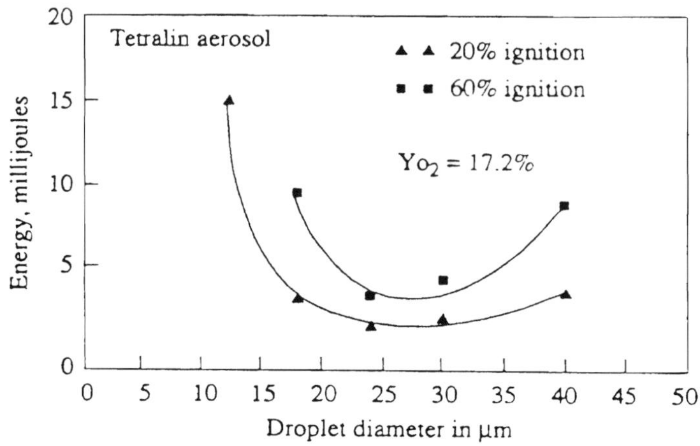

Figure 4.40. Variation of minimum ignition energy of tetralin aerosol with drop size. The mole fraction of oxygen is 17.2%. (Reproduced by permission of Elsevier Science Publishers.)

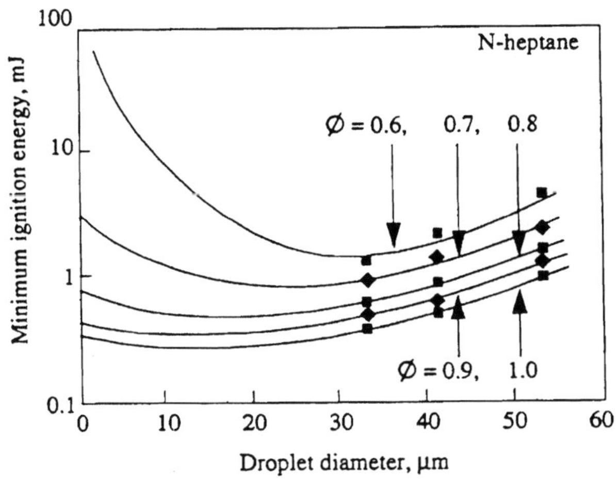

Figure 4.41. Variation of minimum ignition energy of *n*-heptane with drop size and equivalence ratio. (Reproduced by permission of Elsevier Science Publishers)

larger drop sizes. At drop sizes less than about 10 μm it is nearly constant at about the inherent burning velocity of a vapor–air mixture. Mists with larger drop sizes do not fully vaporize even with the passage of a flame front (Burgoyne, 1963; Febo,1995). Thus, not all of the heat of combustion is released in an aerosol explosion.

In experiments with closely controlled drop sizes of tetralin in the range of 7–55 m, Burgoyne and Cohen (1954) found that droplets with diameters below

10 μm behave like a vapor with respect to flammable limits and flame speed. Aerosols with drops above 40 μm behave differently. At low number density, the drops burn individually in their own air envelopes, and one drop ignites the next. As the number density increases, individual droplet flames can no longer be distinguished. Hayashi and Ohtani (1981) observed the same behavior for a nearly monodisperse distribution of *n*-decane. At low number density they saw ignition of individual drops, but no flame propagation through the aerosol, hence nonignition. At higher number density they observed isolated blue-flame (lean) propagation from drop to drop. At still higher number densities there was interacting yellow-flame (rich) propagation.

As a consequence of this burning behavior, the UFL of a mist can be higher than the UFL of a vapor of the same material (Baker et al., 1983, p. 87). The LFL for aerosols is a function not only of concentration but also of mean drop size. In experiments with uniform drop size aerosols generated by flow through a set of hypodermic needles, Cook et al. (1977) showed that the LFL decreases with increasing drop size as illustrated in Figure 4.42. They used the high-boiling tetralin (tetrahydronaphthalene), light hydrocarbon oil OM 13 and heavy oil OM 33 (the numbers indicating the approximate kinematic viscosity in centistokes at 37.5 C). The open circles in Figure 4.42 are obtained by sucking off part of the entrained air used in the aerosol generation so the flammability test would not be influenced by the downdraft of air. They are the more nearly correct values. Figure 4.42 shows a leveling off of the LFL for drop sizes larger than

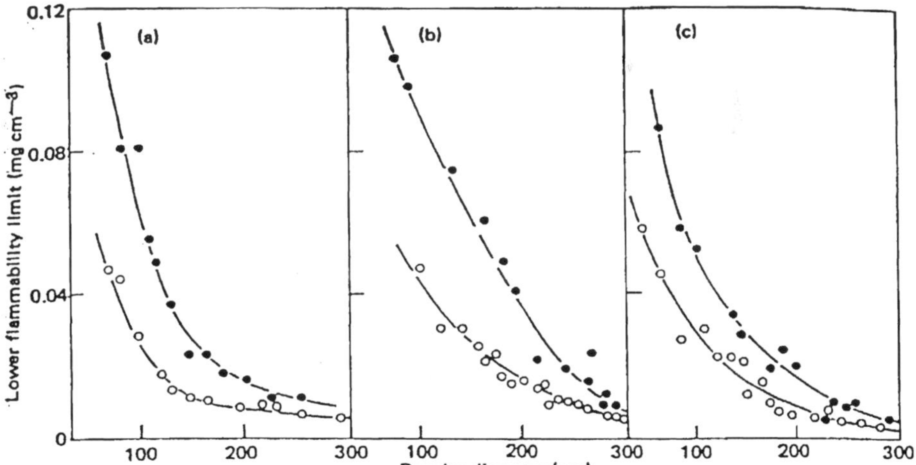

Figure 4.42 The effect of drop size on the lower flammability limit for a monodisperse aerosol of (a) Light OM 13 oil, (b) Heavy OM 33 oil, (c) tetralin: (●) without suction; (O) with suction. (Reproduced by permission of The Combustion Institute.)

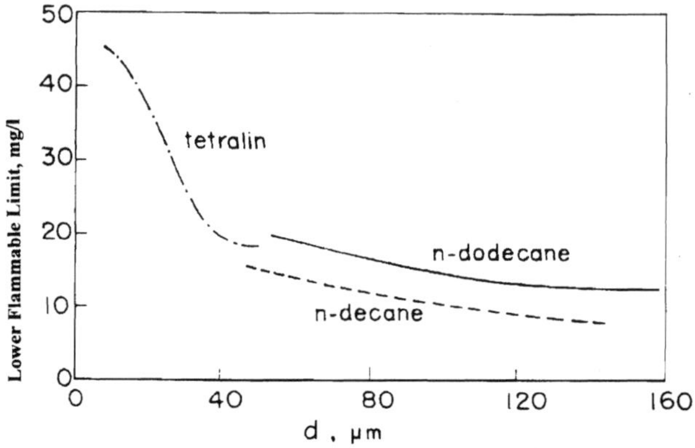

Figure 4.43. Effect of aerosol droplet size on the critical droplet density for ignition of three high-boiling materials. (Reproduced by permission of The Combustion Institute.)

about 200 μm. Drop sizes less than about 100 μm do not readily settle out and usually evaporate totally if not ignited.

Hayashi and Ohtani show a consistent pattern for high molecular weight materials seen in Figure 4.43. The critical drop density, hence aerosol LFL, decreases rapidly to about 40 μm and then decreases more slowly at higher drop sizes.

Burgoyne (1963) pointed out that the settling velocity of droplets adds a downward flux to liquid which effectively increases the concentration seen by an upward flame front. In his measurements he suddenly enclosed a volume of aerosol and measured the liquid mass fraction, w_L, which he denotes as a static mass fraction, w_{Ls}. Let u_d represent the mass-average settling velocity of the drops, and S_t be the upward velocity of the flame front. Then the velocity of the drops relative to the flame is $u_d + S_t$ and the rate of flow of liquid into the flame front, relative to the flow rate of air is a "flame front mass fraction" denoted as a "kinetic" concentration, w_{Lk}, and given by

$$w_{Lk} = w_{Ls}\left(\frac{u_d + S_t}{S_t}\right) \tag{4.71}$$

When the drops are small, the settling velocity is small, and $w_{Lk} \approx w_{Ls}$, but with larger drops, the differences become substantial, as Figure 4.44 illustrates. Figure 4.44 plots measured values of the liquid mass fraction LFL measured by quick capture to obtain a "static" concentration, and by allowing settling to obtain a "kinetic" concentration. The calculated points use Eq. (4.71) to convert from one measurement to a calculated value of the other. This illustrates the importance of knowing how aerosol concentrations are measured in experiments to determine the aerosol LFL and UFL.

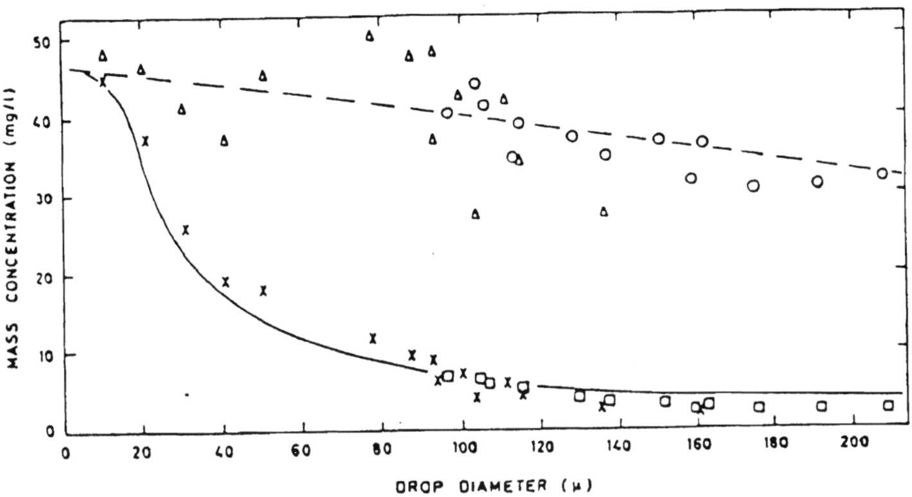

Figure 4.44. Liquid mass fraction measurements in an aerosol by quick capture to obtain a "static" value and by allowing settling to obtain a "kinetic" value of the LFL of tetralin in air (Burgoyne, 1963). (Reproduced by permission of IChemE.)

4.6. Turbulence Effects

To understand how flammable mass is calculated, it is necessary to understand the characteristics of flammable clouds. Concentration profiles in real clouds are not uniform, even though nearly all computer programs predict uniform, mean concentrations in the cross-wind (y,z) plane at any distance (x) downwind. To accurately predict ignition contours we must consider turbulence as well as mean concentrations. That is, the mean concentration at a point may be below the LFL and yet still within an ignition contour because combustible pockets of gas are mixed in by turbulence.

This section discusses experiments and correlations which take turbulence into account for jet plumes and heavy gas plumes.

4.6.1. Turbulence Effects of Jet Plume Ignition

Some general principles of turbulence and ignition are well illustrated with jet plumes and can be applied also to heavy gas plumes. A gas jet in a cross wind is bent by the wind and develops a high degree of turbulence in the under side of the jet. Birch et al. (1989) measured both the concentration profiles and the

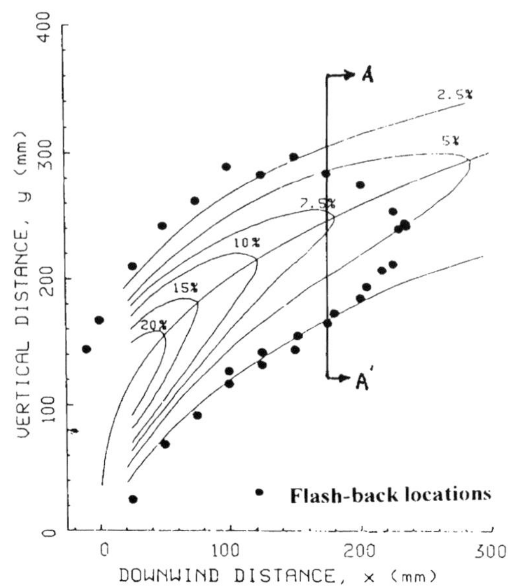

(a)

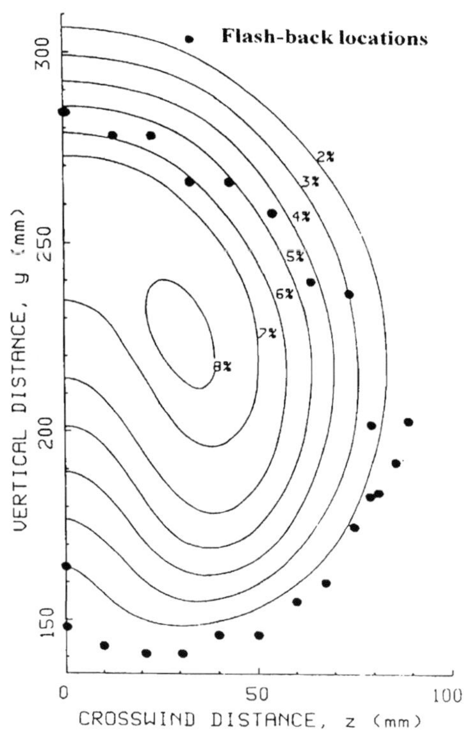

(b)

Figure 4.45. Contours of mean concentration and flash-back ignition points on the symmetry plane of a methane jet. (Reproduced by permission of Gordon and Breach Science Publishers.)

ignition contours of a methane jet as shown in Figure 4.45. The solid dots outline an ignition contour, within which a flame flashes-back upon ignition. Outside this contour a flame is advected downstream and quenched. The side-view of a vertical jet in a cross-wind, Figure 4.45a shows that the ignition contour corresponds to about the 2.5% concentration contour, or ½ of the LFL. Figure 4.45b plots the right-half contours on a vertical plane through the symmetrical jet. The concentration profiles in the vertical plane develop a characteristic kidney-shape because the drag on the plume pulls down the edges of the plume in a double vortex. The ignition contour falls along the LFL on the top edge of the jet plume where there is little turbulence. It falls near the ½ LFL on the underside of the jet where the turbulence is highest and turbulence produces local pockets of concentration above the LFL.

In earlier work, Birch et al. (1981) provide turbulence profiles across a jet of methane. The data of Birch et al. were replotted as ignition probability as a

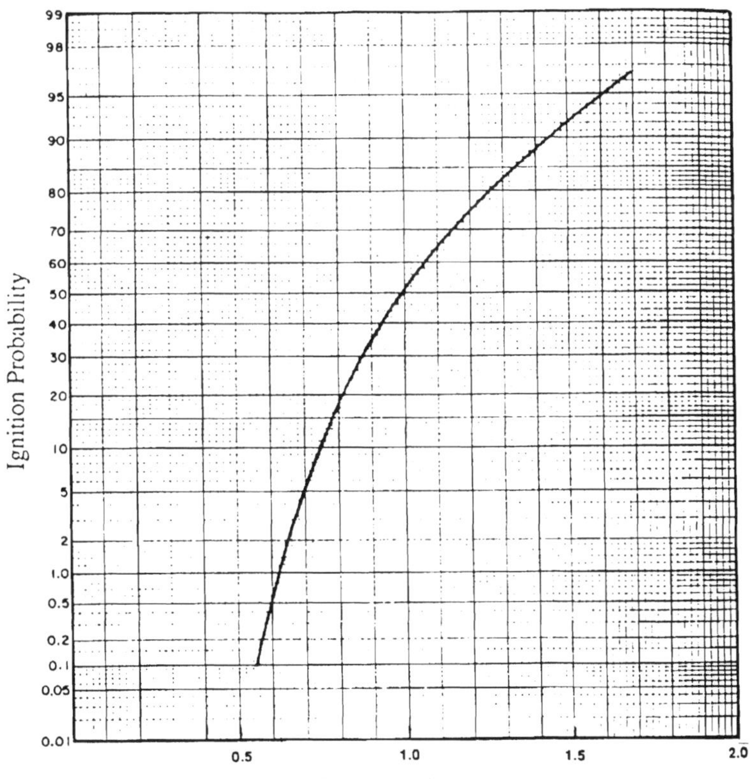

Figure 4.46. Ignition probabilities of a natural gas jet.

function of mean concentration, centering about the LFL in Figure 4.46 (Gustafson and Mudan, 1987). Their curve shows that for a jet the probability of ignition is 10% at 0.75 LFL, 50% at the LFL, and 95% at 1.6 the LFL.

4.6.2. Turbulence and Pockets of Flammable Material

Normally, dispersion models produce concentration profiles for mean values of concentration at any point (x,y,z). These concentration profiles are not entirely coincident with ignition profiles because the probability of ignition is a function of the turbulence profiles in a plume, as discussed in the previous subsection. We next show some correlations that can be used to relate turbulence and the probability of ignition to mean cloud profiles.

Csanady (1969) studied concentration fluctuations for heavy gas clouds over various terrains and observed that the turbulent fluctuation intensity depends on terrain and surface roughness. The cumulative probability, P, of exceeding a limiting concentration, c_{lim}, such as ½ LFL given a mean concentration, $\bar{c}$, and a standard deviation, σ_{p}, is expressed by

$$\frac{\bar{c}}{c_{\text{Lim}}} = \exp\left\{-2\sqrt{\sigma_{\text{p}}}\,\exp[-(1-2P)] + \frac{\sigma_{\text{p}}^2}{2}\right\} \tag{4.72}$$

Csanady provides the standard deviations for a logarithmic normal distribution of turbulence in Table 4.8. With these values, probabilities are plotted for exceeding a limiting concentration such as the LFL in Figure 4.47 (Csanady 1969).

Spills of cryogenic liquids such as LNG and LPG on a river or bay will rapidly evaporate and entrain air doing so. They generate a different turbulence spectrum than occurs with heavy vapor clouds. Fay and Lewis (1975) analyzed concentration fluctuations from nearly instantaneous spills of liquefied natural gas (LNG 95% methane) on water. Their results in Figure 4.48 plot the probability of ignition as a function of mean concentration of hydrocarbon. The probability of ignition at the LFL of 5.0 % is about 40%. At ½ LFL (2.5 %) the probability

TABLE 4.8
Logarithmic Standard Deviation of Concentration Fluctuations in Vapor Clouds

Weather Condition	Terrain Condition	Standard Deviation
Stable (Pasquill stability class E and F)	Smooth unobstructed terrain	0.2
Stable (Pasquill stability class E and F)	Rough terrain	0.35
Neutral (Pasquill stability class C and D)	Smooth terrain	0.35
Neutral to unstable	Rough terrain	0.70

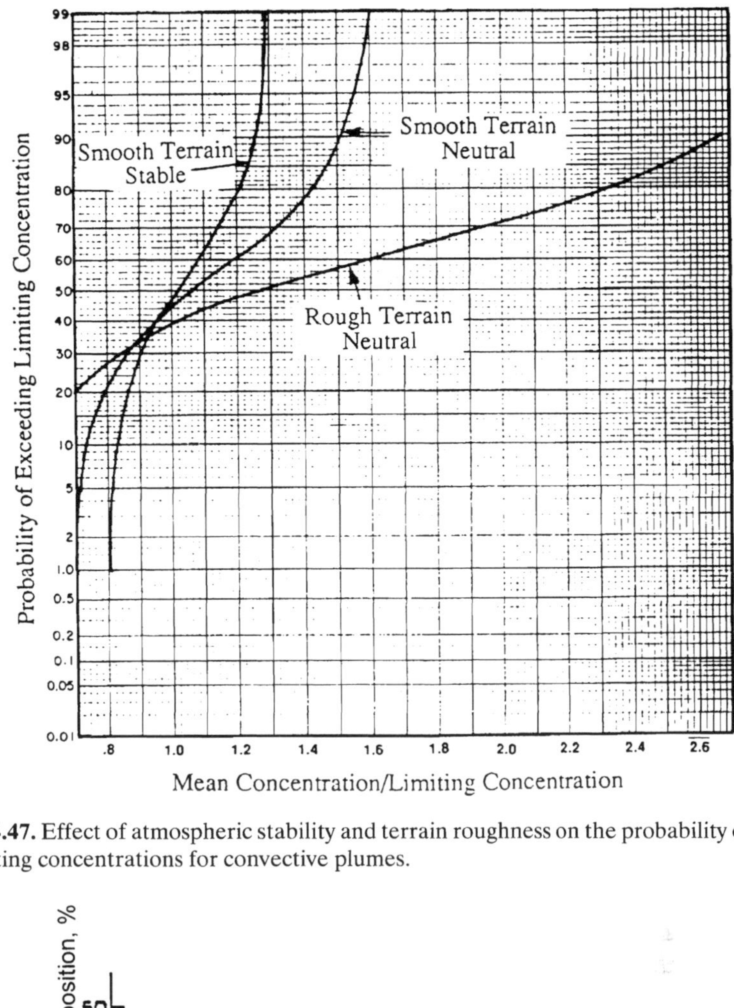

Figure 4.47. Effect of atmospheric stability and terrain roughness on the probability of exceeding limiting concentrations for convective plumes.

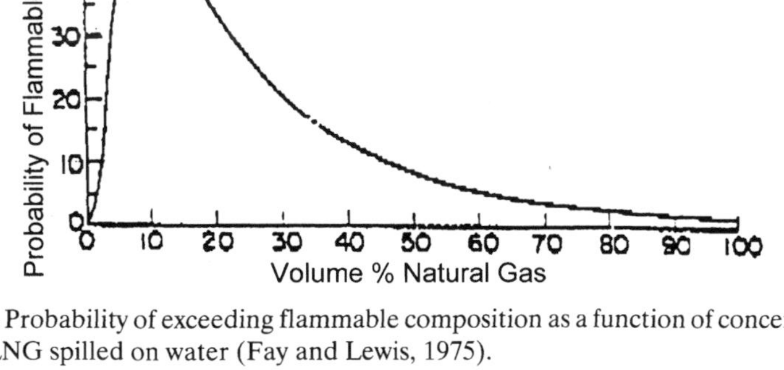

Figure 4.48. Probability of exceeding flammable composition as a function of concentration for plumes of LNG spilled on water (Fay and Lewis, 1975).

of ignition is about 10%. By comparison with Figure 4.46, a jet produces less turbulent "broadening" than a cloud formed by a boiling liquid. (The 10% probability of ignition occurs at 0.75LFL for a jet and at 0.5LFL for a boiling cloud.) For conservative estimates, an ignition probability of 10% is often used.

4.7. Flash Fires

If a flammable cloud ignites, but does not explode, the initial damage will be primarily by thermal radiation. However, flash fires may generate more-damaging "knock on" events, especially if they burn back to the source. The knock-on event could be one of the following:

- Jet fire
- Pool fire
- BLEVE
- Partially confined vapor cloud explosion
- Confined vapor cloud explosion

The first step to model a flash fire is to predict the flammable mass. In order to evaluate the distance traveled by the flammable cloud before ignition, some estimate is usually made of the probability of ignition (Clay et al, 1988). The boundary of the fire can be approximated by the unignited cloud's downwind and crosswind dimensions at flammable limit concentrations (LFL or ½ LFL).

Flash fire models are available to model the ignited cloud case (see CCPS, *Guidelines for Evaluating the Characteristics of Vapor Cloud Explosions, Flash Fires, and BLEVEs*). With these models, fire radiation contours can be drawn to define hazard zones. View factors are considered in drawing radiation contours in this case. For example, the model of Considine et al. (1982) is for diffusion fires with an equivalence ratio, $\phi > 1$, and is similar to that of Raj and Emmons (1975). The flame front is taken as a wall of flame of specified thickness and height, traveling back through the cloud with a turbulent burning velocity.

4.8. Explosions

The types of explosions usually defined are condensed-phase (solid or liquid state) explosions, phase-change explosions, pressure rupture, vapor cloud explosions, and dust explosions. TNT and plastic explosive are typical condensed-phase explosives. Phase-change explosions occur, for example, when molten slag is inadvertently poured into a rail car containing a thin pool of water. Some damaging explosions have occurred in tests spilling liquefied natural gas (LNG) that were phase-change explosions since no combustion took place. Boiler explosions typify pressure ruptures. Dusts typically have very high flammable limits, so much so that a dust explosion is usually a secondary, knock-on event. A smaller disturbance or explosion will dislodge dust settled on overhead beams, etc. and generate the high dust concentrations needed for a dust explosion. For this reason dust cloud concentrations are not readily amenable to calculating the flammable mass prior to ignition.

Here, we treat only vapor cloud explosions. It is not our purpose here to provide an in-depth discussion of explosion modeling, but rather to briefly outline a few principles to demonstrate the application of flammable mass prediction. For more details, see definitive works by Nettleton (1987), Baker (1973), and the CCPS book *Guidelines for Evaluating the Characteristics of Vapor Cloud Explosions, Flash Fires, and BLEVEs*.

Vapor cloud explosions, to generalize the term, occur also with liquid aerosol clouds, as discussed in Section 4.5. In addition, they occur with solid–liquid mixtures such as propylene–polypropylene, ethylene–polyethylene, and propylene–polypropylene releases (Cardillo, 1980; Tamanini, 1996). The latter solid–flammable-vapor systems, referred to as hybrid systems, have flammable limits and other flammability properties similar to vapors and aerosols.

4.8.1. Confinement and Congestion

Experiments made with vapor clouds in a plastic tent or a balloon in the absence of congestion showed that explosions generally do not occur, only flash fires (Leyer and Desbordes, 1993; Lind, 1975). Other experiments show that congestion and/or confinement are usually needed for flame acceleration to explosive levels. An increase in congestion and in the size of the area will, in most cases, increase explosion overpressures. An exception is when the added congestion acts to hinder gas flow. Repetitive flow pattern disruption is believed to be necessary to initiate the postive feedback mechanism that increases flame speeds exponentially.

Confinement is defined by the presence of physical surfaces which significantly limit flame expansion. Figure 4.49 sketches the degrees of confinement, or as used here, the dimensions available for flame expansion. A zone is considered three dimensional (3D) if the flame is free to expand in all directions. It is two dimensional (2D) if the flame is free to expand in two dimensions (an area with open sides or a solid deck above). It is one dimensional (1D) if the flame can expand in only one direction (a pipe, culvert, or narrow tunnel). The Baker–Strelow method (Baker et al., 1994) recommends that multideck open-grating structures be defined as 2D. Confinement is further developed in Section 4.8.6.

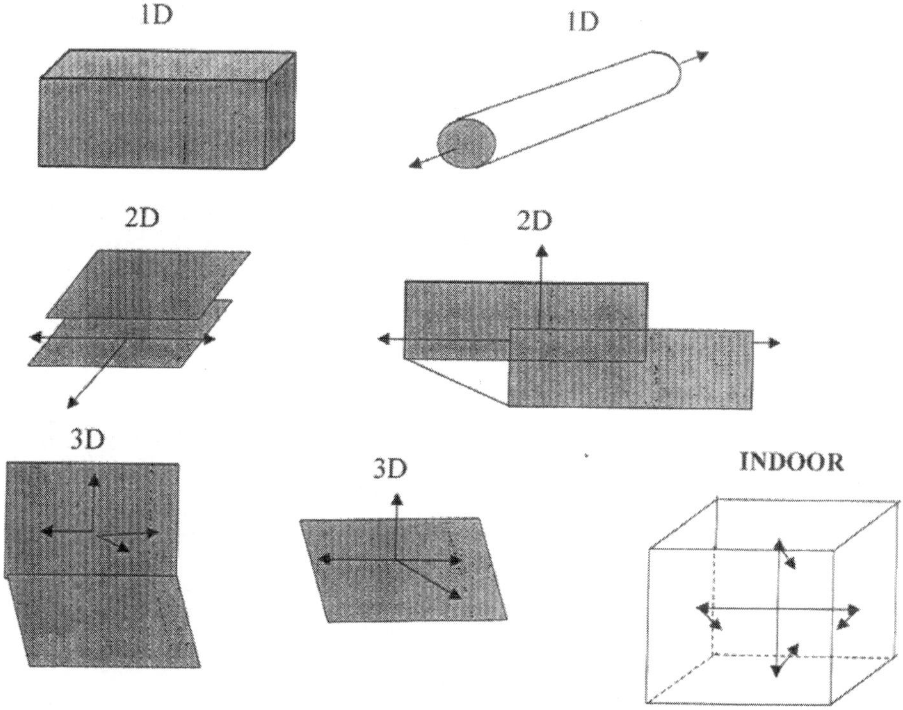

Figure 4.49. Examples of flame expansion geometries.

If the confinement is increased to the point of total enclosure, an explosion is much more likely than with partially confined vapor clouds. Confined flammable mass and confined explosions are discussed in Section 5.4.

Congestion is defined by the presence of obstacles of a size that allow fluids or flame fronts to readily flow by, but in doing so the flow generates turbulence. Turbulence can act to accelerate flame speeds. Characterizing the degree of congestion for plant conditions, is not a straightforward process. Experiments on the effect of congestion have used simplified geometries such as that shown in Figure 4.50 (CMR, 1993, p. 63). Bjerketvedt et al. (1993) performed experiments in a corner with two walls using several combinations of pipe diameters and volume blockage ratios (VBR) defined as the volume of the obstacles divided by the volume bounded by the obstacles.

In their experiments with a stoichiometric propane-air mixture the flame speed ranged from 10 m/s without obstacles to approximately 1000 m/s in the most densely packed arrangement. The pressure produced ranged from a few

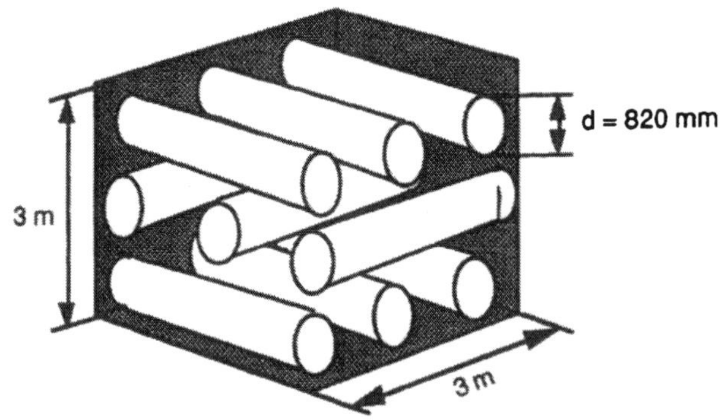

Figure 4.50. Defining degree of congestion by volume area blockage and pipe diameter. (Reproduced by permission of Christian Michelsen Research AS.)

kPa up to 400 kPa (4 barg). Figure 4.51 (CMR, 1993, p. 63) shows the experimental results for the combinations of diameters and VBR. Other measures of congestion are discussed in Section 4.8.7.

When an indoor vapor cloud fills several rooms that are connected by corridors or a hallway, an explosion can amplify from room-to-room by a phenomenon named precompression or pressure piling (Crowl & Louvar, 1990, p. 373). The expansion wave from the explosion raises the pressure in "downstream" rooms so when the combustion wave arrives it reacts at a higher pressure and produces even more damaging overpressures than otherwise. Few data are available for this situation.

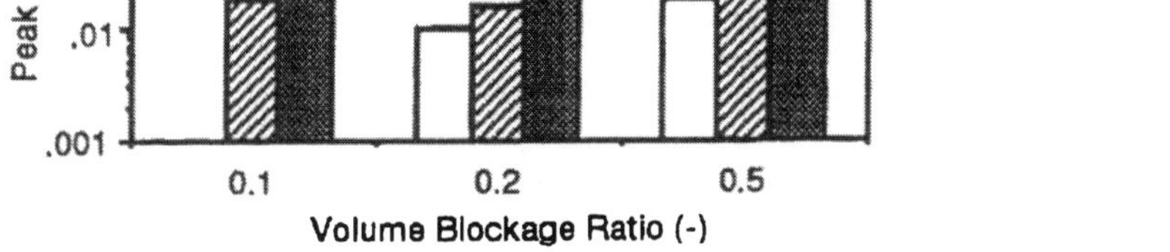

Figure 4.51. Experimental peak overpressures as a function of volume blockage ratios for stoichiometric propane–air in a 27 m^3 cubical explosion vessel. (Reproduced by permission of Christian Michelsen Research AS.)

A large gas cloud may cover two process areas located near to each other. An explosion starting in one area may progress into the second before the turbulence has decayed and the flame acceleration in the other area could increase rapidly creating a higher pressure than for the first area.

If the gas cloud extends beyond a congested area the flame acceleration will decay when the end of the congestion is reached. Figure 4.52 shows the results of an experiment in which two areas of congestion, each 2 m wide, are separated by an open, uncongested area of width $S = 0.5$, 1, and 2 m (Mercx, 1992, 1998). The flame speed accelerates in the first congested area, decelerates in the open area, and then accelerates again in the second congested area. In general, the rate of the decay in flame speed upon leaving the congested area is dependent on the size of the equipment or objects creating the turbulence. If mostly small sized piping is creating the turbulence, the decay is rapid and the flame speed may reach a low level within a few meters. Otherwise, more than 3–5 m may be needed to slow down the flame and overpressures to nondamaging levels.

Wind passing through equipment in the plant upstream of the release source will increase the turbulence level. However, such contributions to turbulence are in most cases smaller than that obtained from a large gas jet leak.

4.8.2. Effect of Concentration on Explosion Overpressure

Peak overpressures in an explosion also depend upon concentration as illustrated in Figure 4.53. This plot, from Vanderstraeten et al. (1997) shows that the maximum overpressure for a methane explosion in a confined vessel is lowest near the flammable limits (5% and 15%) and is highest near the stoichiometric concentration (9.5%). Dimensionless explosion overpressure, $\Delta P/P_0$, is directly

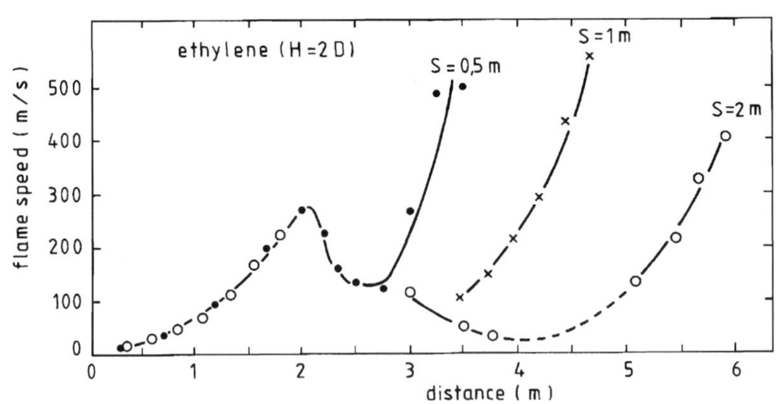

Figure 4.52. Experimental flame speeds in 2 m wide congested regions separated by an uncongested gap of width *S*. (Reproduced by permission of TNO.)

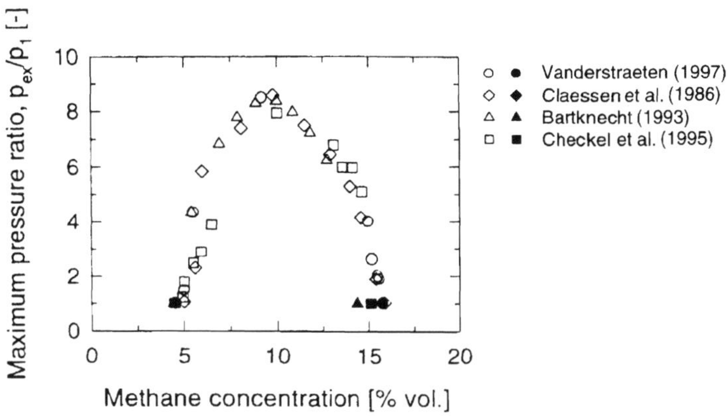

Figure 4.53. Maximum explosion pressure ratio for methane/air mixtures ignited at 20 C with P_1 = 100 kPa (Reproduced by permission of Elsevier Science Publishing Co.)

related to the square of flame speed (as Mach Number, $M_x = u_x/a$), and the isentropic coefficient, γ (Baker, 1973, p. 154):

$$\frac{\Delta P}{P_0} = \frac{2\gamma(M_x^2 - 1)}{\gamma + 1} \tag{4.73}$$

Flame speed is seen to vary with structural confinement, turbulence-inducing congestion, and concentration gradients, which differs from one event to another. One reason why the explosion efficiency for TNT equivalence models (defined in the next section) varies for different events is explained by Figure 4.53.

4.8.3. TNT Equivalence Explosion Models

Taking advantage of the fact that TNT explosions have been well studied, TNT equivalence models have been in common use to calculate vapor cloud explosion characteristics. Over the years a number of slightly different TNT equivalence curves have been published.

When explosion experiments are conducted, the following input values are known

- Ambient pressure, P_0 (Pa)
- Radial distance from the explosion, R (m)
- Mass of explosive or equivalent mass of TNT, W (kg)
- Combustion energy released in the explosion, E (J), or, for TNT, the explosion energy
- Speed of sound at ambient conditions, a_0 (340 m/s at 293 K and 1 atm)

The experiments measure the following values:

- Peak overpressure, P (Pa)
- Peak arrival time, t_a (s)
- Impulse, I (integral of overpressure as a function of time) (Pa.s)
- Pulse duration, t_d (s)
- Negative peak and duration

To generalize the experiments, the data are transformed to scaled variables. For the TNT equivalence models, scaling is based on an equivalent mass of TNT, W. This mass is found by equating the energy release in a vapor cloud explosion to the energy release in a TNT explosion. The energy in a vapor cloud is the flammable mass, m_f, times the net heat of combustion, H_c, times a yield, or efficiency factor:

$$E = \varepsilon m_f H_c \tag{4.74}$$

$$E_{TNT} = W H_{TNT} \tag{4.75}$$

Here, the flammable mass is the mass of fuel between the LFL and the UFL. The energy in the TNT used in a particular experiment is similarly the mass of TNT, W, times the detonation energy of TNT, H_{TNT}. A commonly used value for H_{TNT} is 4.648×10^6 J/kg. Alkane hydrocarbons like ethane, propane, butane etc., have a net heat of combustion, H_c, about 10 times the detonation energy of TNT. Equating the two energy terms and solving for W gives

$$W = \varepsilon m_f \frac{H_c}{H_{TNT}} \tag{4.76}$$

Figure 4.54A,B from Kingery and Bulmash (1985) published in Lees (1995) gives TNT equivalence explosion curves which clearly distinguish between elevated free air bursts (with spherical symmetry) and for ground-level bursts (with hemispherical symmetry). The ground-level burst overpressure curves are higher than the corresponding free air burst curves. In texts in which only the free air burst curves are available, for cases when the blast is from a ground-level source, the user is advised to double the effective energy of the explosion (or the mass of fuel contributing to the explosion). This correction accounts for the approximate doubling of momentum from ground-level reflection. This effect is already built in to ground-level curves.

The curves in Figure 4.54 make use of TNT mass scaling, and have dimensions such as m/kg of $TNT^{1/3}$. Similar curves for the volume-based explosions models discussed in the next section use Sachs (1944) scaling parameters and result in dimensionless scaled variables. These scaling rules are defined in Table 4.9.

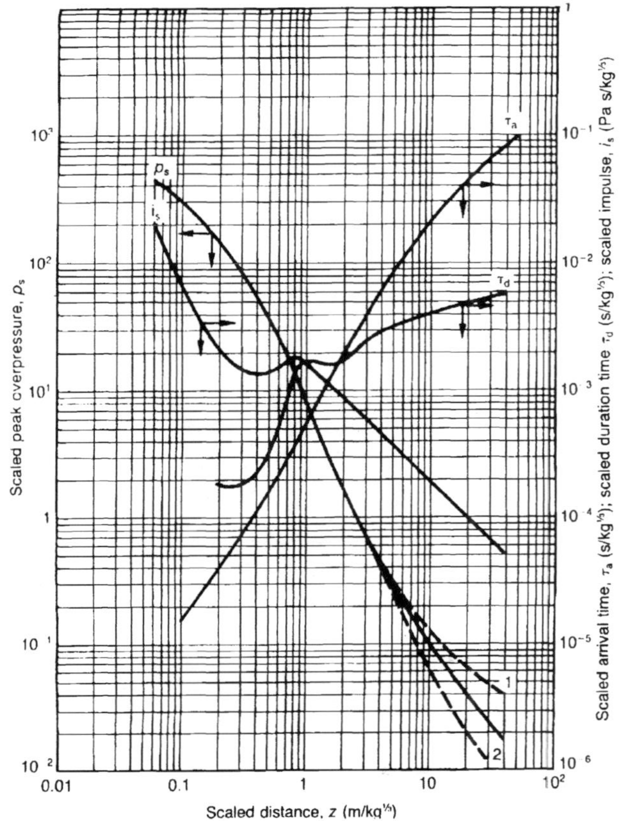

Figure 4.54A. Relative peak side-on overpressure, impulse, arrival time, and completion time for a free air burst of TNT (spherical symmetry) from Kingery and Bulmash, 1985 (from p 17/131 of Lees, 1996). (Reproduced by permission of Butterworth-Heinemann.)

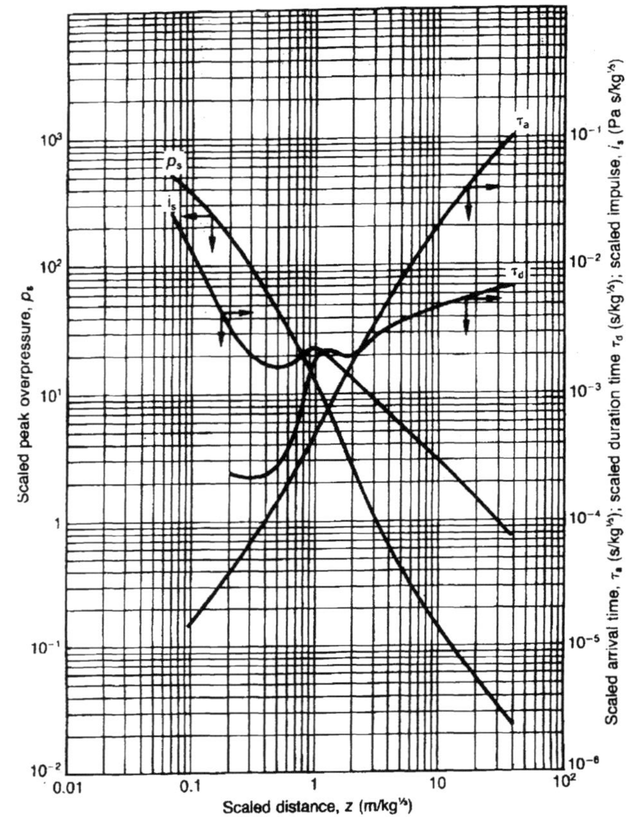

Figure 4.54B. Relative peak side-on overpressure, impulse, arrival time, and completion time for a surface burst of TNT (hemispherical symmetry) from Kingery and Bulmash, 1985 (from p 17/131 of Lees, 1996). (Reproduced by permission of Butterworth-Heinemann.)

TABLE 4.9
Scaling Rules for Explosion Models

Scaled Variable	Energy Scaling (for TNT Equivalence Models)	Sachs Scaling (for Volume-Based Models)
Radial "Stand-Off" Distance, Z	$Z = \frac{R}{W^{1/3}}$	$Z = \frac{R}{(W/P_0)^{1/3}}$
Peak overpressure, P_s	$P_s = \frac{\Delta P}{P_0}$	$P_s = \frac{\Delta P}{P_0}$
Impulse, i_s	$i_s = \frac{I}{W^{1/3}}$	$i_s = \frac{I a_0}{E^{1/3} P_0^{2/3}}$
Impulse duration, τ_d	$\tau_d = \frac{t_d}{W^{1/3}}$	$\tau_d = \frac{t_d a_0}{(E/P_0)^{1/3}}$
Arrival time, τ_a	$\tau_a = \frac{t_a}{W^{1/3}}$	$\tau_a = \frac{t_a}{R/a_0}$

The TNT equivalence method reduces the problem of vapor cloud explosion modeling to one of finding a valid value for explosion efficiency. Explosion efficiency is found from industrial accidents by using damage correlations with overpressure to find the term m_{TNT}. Then estimate the flammable mass in the cloud and solve for ε. Unfortunately ε can be found using either the flammable mass in the vapor cloud, m_f, or of the total mass released, m_t. Considerable confusion could be avoided by defining the former as a net efficiency, ε_{net} and the latter as a gross efficiency, ε_{gross}. The relationship between these terms is

$$\varepsilon_{gross} = \varepsilon_{net} f_m \tag{4.77}$$

where f_m = fraction of released material in the flammable region = m_f/m_t

Since f_m could well range to 10% or less, a gross efficiency would be 10% or less of the net efficiency. To erroneously apply the gross efficiency times a calculated flammable mass that already includes the f_m factor would give an explosion energy that is signficantly low.

A histogram compiling explosion efficiency is given in Figure 4.55 by Shepherd et al. (1991). This histogram shows a wide range for explosion efficiency, from 0.1% to nearly 100%. A possible reason for the wide spread is that the compilation could contain both net and gross explosion efficiencies. As the estimates of flammable mass released improve, and net explosion efficiencies are calculated instead of gross efficiencies, the range of explosion efficiencies is expected to narrow.

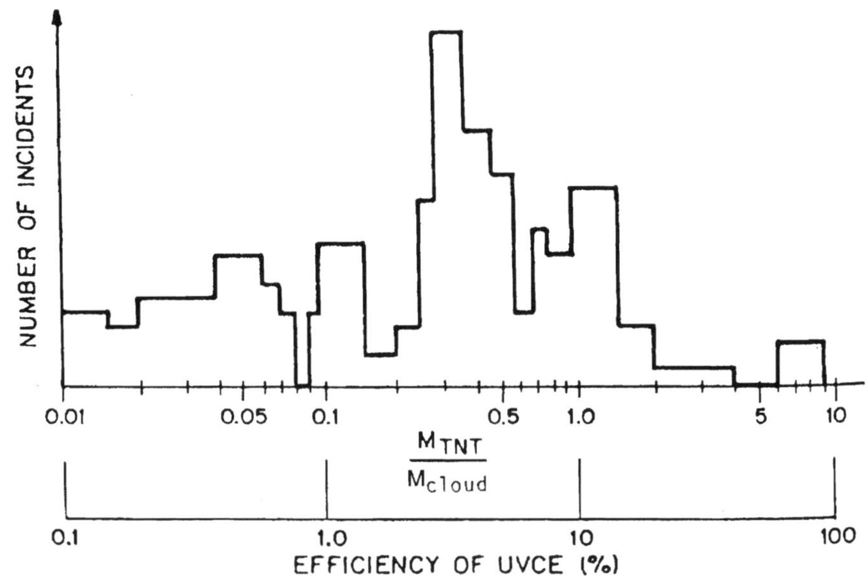

Figure 4.55. TNT equivalence histogram of unconfined vapor cloud exposions (Shepherd et al., 1991).

For risk analyses, historically, gross explosion efficiencies of 3 – 5% are typically used. Gross efficiencies of 10% to 15% would be suitable for more reactive materials (*Yellow Book*, 1997, pp. 5.23–5.24; *Guidelines for Evaluating the Characteristics of Vapor Cloud Explosions, Flash Fires, and BLEVEs*, 1994). Net efficiencies would be 3 to 5 times higher.

To use Figure 4.54, find values for the input terms, P_0, E, W, and a and specify values of interest for the "stand-off" distance R. At each R calculate the dimensionless value, Z, and look up from Figure 4.54 the values P_s, τ_a, and I_s or τ_d.

Example 7 Use of TNT Equivalence Curves

Consider a release of ethylene with a flammable mass of 575 kg. From Appendix C, the net heat of combustion of ethylene is 47.15 MJ/kg. Taking a net explosion efficiency based on the flammable mass of 12%, the equivalent mass of TNT is

$$W = \frac{0.12 * 575 * 47.15E6}{4.648E6} = 700 \text{ kg TNT}$$

The scaling factor, $W^{1/3}$ is 8.88 $kg^{1/3}$. Tabulated on the following page are the values read from the scaled plots, Figure 4.54A,B, for a free-air burst and a ground-level burst at scaled distances of $Z = 1$ and 10. The scaled values are translated to unscaled values in Columns 5 and 6. Note that the ground-level burst produces a larger overpressure, impulse, and impulse duration than the free-air burst.

Scaled Values			Unscaled Values		
Z (arbitrary)	1	10	Z	1	10
R (m)	8.88	88.8	R (m)	8.88	88.8
			Free-Air Burst		
P_s (–)	7.88	0.10	P (bar)	7.88	0.10
i_s (Pa.s/$kg^{1/3}$)	1.85E-3	0.19E-3	I (Pa ms)	16.4	1.7
τ_d (s/$kg^{1/3}$)	1.9E-3	4.0E-3	t_d (ms)	16.9	35.5
τ_a (s/$kg^{1/3}$)	0.66E-3	23.5E-3	t_a (ms)	5.9	208
			Ground-Level Burst		
P_s (-)	12.3	0.14	P (bar)	12.3	0.14
i_s (Pa.s/$kg^{1/3}$)	2.29E-3	0.304E-3	I (Pa ms)	20.3	2.7
τ_d (s/$kg^{1/3}$)	1.97E-3	4.70E-3	t_d (ms)	17.5	42
τ_a (s/$kg^{1/3}$)	0.50E-3	20.0E-3	t_a (ms)	4.4	178

4.8.4. Volume Source Explosion Models

Vapor cloud explosions differ significantly from TNT explosions. As illustrated in Figure 4.56 (Marshall, 1976; Lees. 1996, p. 17-161), in a TNT explosion the highest pressure is at the ignition point, and pressures decay outward. In a vapor cloud explosion, the overpressure builds to a maximum as the flame accelerates from the ignition point. Thus, a maximum overpressure develops some distance from the point of ignition. In addition, the maximum overpressure observed from a partially confined vapor cloud explosions is about 1 bar compared with about 400 bar for TNT explosions. Thus, portions of the TNT equivalence curves in Figure 4.54 very near the center of the explosion are inapplicable to vapor cloud explosions. In addition, TNT equivalence curves tend to overpredict damage levels in the far field and underpredict in the near field. Furthermore, the TNT equivalence explosion model does not account for the effects of congestion and confinement.

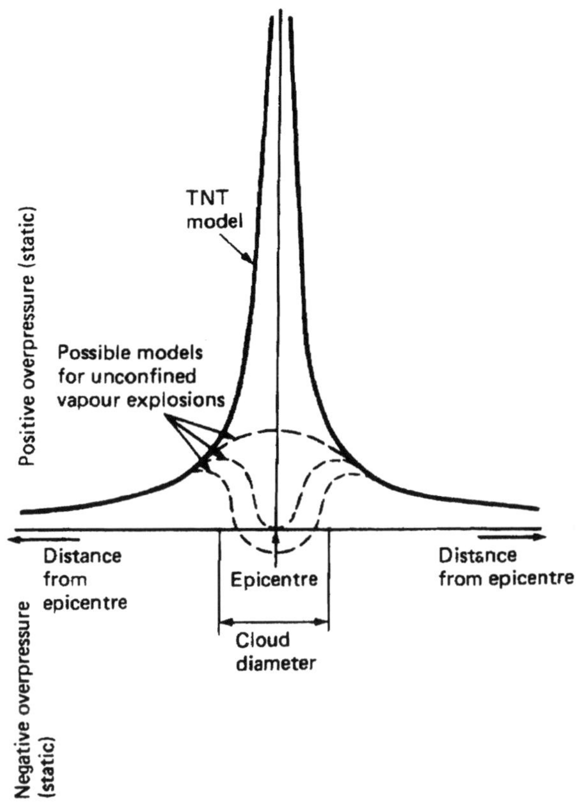

Figure 4.56. Conceptual comparison of explosion overpressures from TNT and vapor cloud explosions.

To more accurately represent vapor cloud explosions, and to account for the effects of partial confinement and congestion, volume-source explosion models have been developed. Two such models are used in examples in Chapter 7, the Baker–Strehlow model (Baker et al., 1994, 1997) and the TNO multi-energy method (CCPS, 1994; van den Berg, 1985).

The basic principles of the multi-energy method and the Baker–Strehlow method for explosion modeling are that areas of confinement and/or congestion are the basis for determining the mass of the flammable vapor cloud that contributes to the explosion. During a deflagration, flames accelerate while traveling through congested zones and decelerate after exiting such a zone. Damaging blast waves are produced in the confined/congested areas, whereas a fireball with negligible overpressure occurs in an unconfined/uncongested area. As a result, only the energy in fuel within congested/confined zones is taken as the explosion energy. Multiple blast sources can emanate from a single release if that release passes through multiple congested areas. The flammable mass in each congested/confined zone must be calculated separately since each generates a separate event, although separated by milliseconds.

The explosion overpressure and impulse curves for the Baker–Strehlow model are shown in Figure 4.57A,B and for the multi-energy method in Figure 4.58A,B. Both use Sachs scaling defined in Table 4.9.

The multi-energy and Baker–Strehlow methods are essentially equivalent in all important respects. On each figure they both plot several curves which depend upon a maximum overpressure parameter. These parameter values are correlated with degree of congestion and confinement. Confinement is expressed as the dimensions available for flame expansion as illustrated in Figure 4.49. The Baker–Strehlow method uses as its maximum overpressure parameter the maximum turbulent flame speed expressed as a Mach number, M_w, and summarized in Table 4.10.

The multi-energy method uses values from 1 to 10 as its maximum overpressure parameter. Based on a review of experimental data Kinsella (1993) related the maximum overpressure parameters to the dimensions available for flame expansion, obstacle density, and ignition energy in Table 4.11A. High ignition energies correspond to "hard" ignition sources listed in Section 4.3 and represent energies that almost immediately push flame speeds to a detonation. These high energy levels are a function of the gas reactivity, since less reactive

TABLE 4.10
Parameter on Baker–Strehlow Explosion Overpressure/Impulse Curves as Turbulent Flame Speed in Mach Number (M_w)

		Obstacle Density		
Flame Expansion	Fuel Reactivity	High	Medium	Low
1D	High	5.2	5.2	5.2
	Medium	2.27	1.77	1.03
	Low	2.27	1.03	0.294
2D	High	1.77	1.03	0.588
	Medium	1.24	0.662	0.118
	Low	0.662	0.471	0.079
3D	High	0.588	0.153	0.071
	Medium	0.206	0.100	0.037
	Low	0.147	0.100	0.037

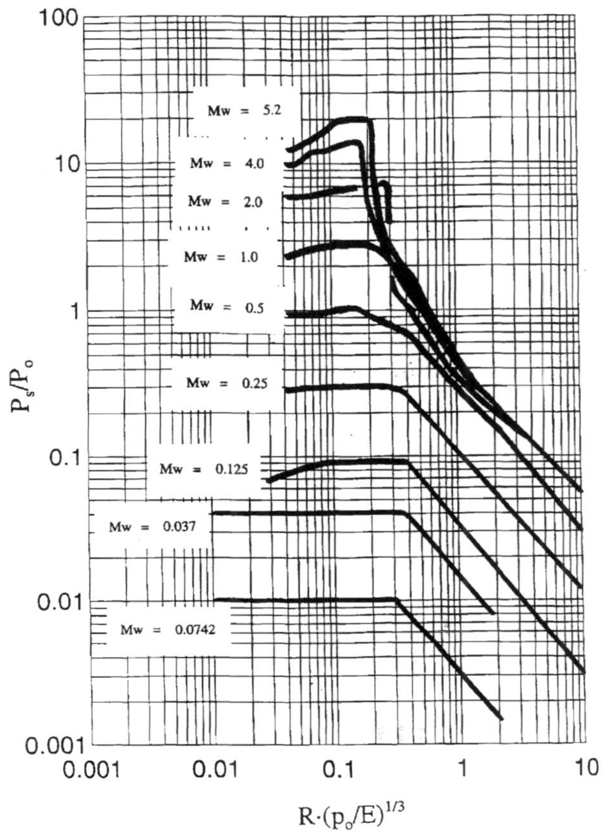

Figure 4.57A. Scaled overpressure for partially confined vapor cloud explosion by the Baker–Strehlow method (Baker et al., 1994, 1997).

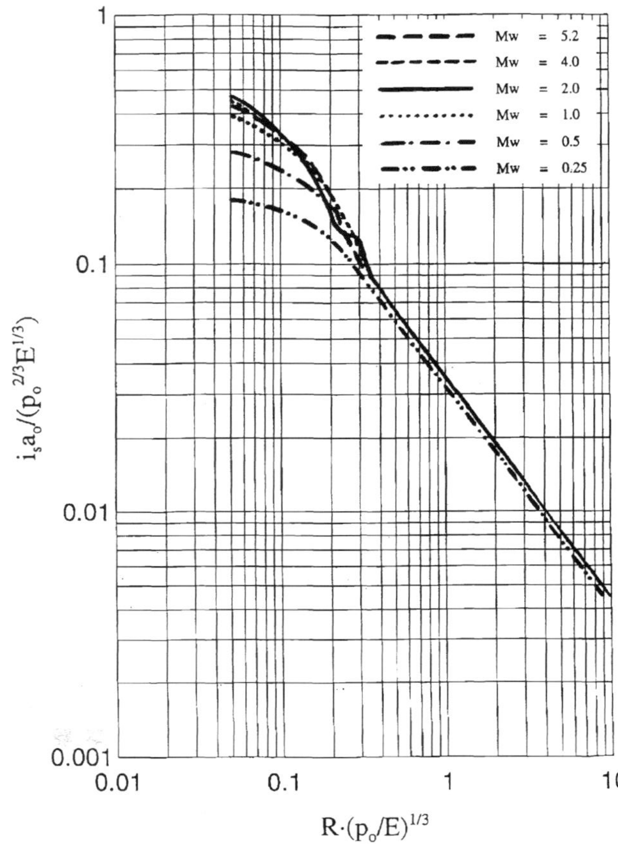

Figure 4.57B. Scaled impulse for partially confined vapor cloud explosion by the Baker–Strehlow method (Baker et al., 1994, 1997).

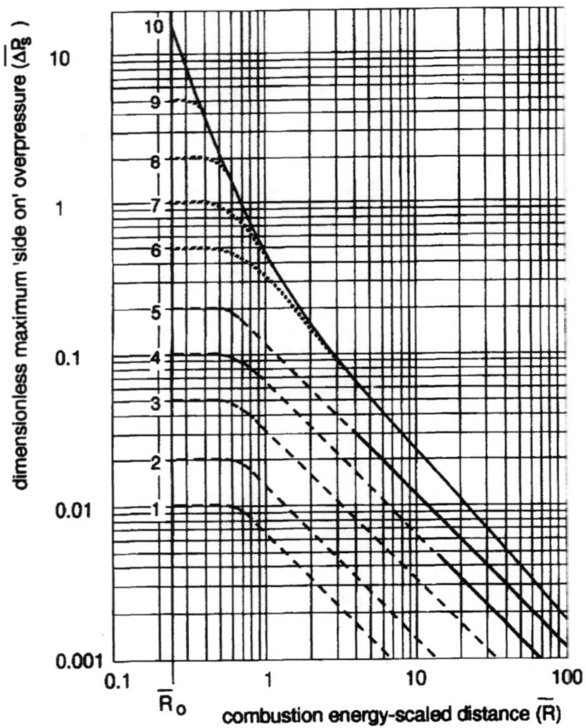

Figure 4.58A. Scaled overpressure for partially confined vapor cloud explosion given by the multi-energy method. (Reproduced by permission of Elsevier Science Publishers, Inc.)

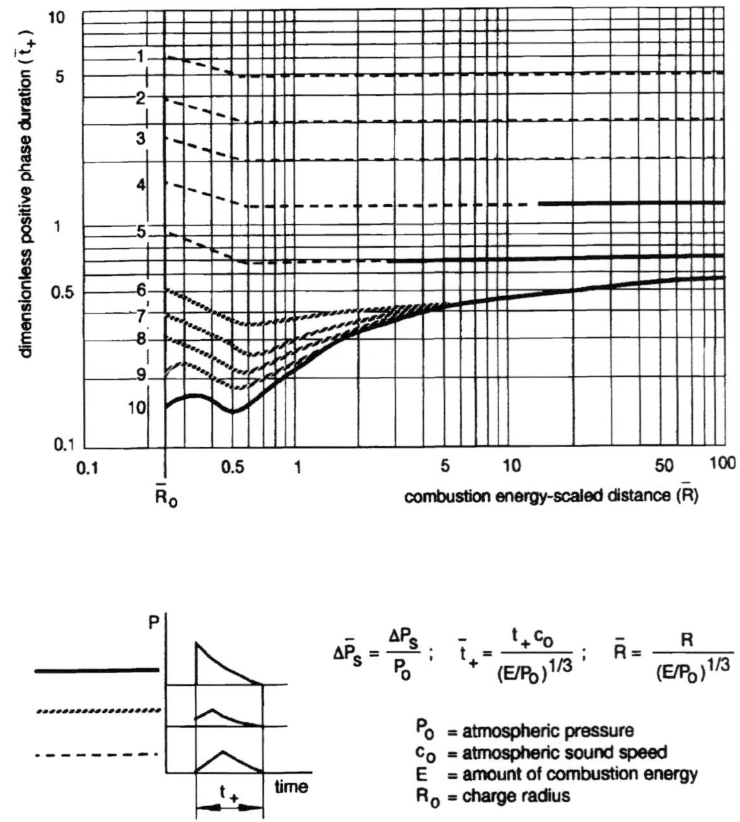

Figure 4.58B. Scaled impulse curves for partially confined vapor cloud explosion by the Multi-energy method. (Reproduced by permission of Elsevier Science Publishers, Inc.)

materials such as methane require a larger ignition energy to produce a detonation. Experiments give the high ignition energies in terms of mass of explosive. Translating to explosive energy gives values for high ignition energy of around 1J. The corresponding maximum overpressures are listed in Table 4.11B. An alternative correlation for the maximum overpressure parameter based on degree of congestion is discussed in Section 4.8.6.

The methods differ slightly in their treatment of scaled impulse. The Baker–Strehlow method plots scaled impulse directly in Figure 4.57B. The multi-energy method plots the dimensionless pulse duration in Figure 4.58B, which is proportional to scaled impulse pressure. Assuming a triangular-shaped pressure pulse, the impulse is given by the peak overpressure at the target structure, ΔP and the duration of the impulse, t_d, which can be found from the corresponding scaled values:

$$I = \frac{1}{2}\Delta P t_d \tag{4.78}$$

TABLE 4.11A
Multi-Energy Method Explosion Model Parameter Definition by Kinsella (1993)

Flame Expansion	Ignition Energy	Obstacle Density		
		High	Low	None
2D	High	7 – 10	5 – 7	4 – 6
	Low	5 – 7	3 – 5	1 – 2
3D	High	7 – 10	4 – 6	4 – 5
	Low	4 – 5	2 – 3	1

TABLE 4.11B
Definition of Multi-Energy Method Peak Overpressure Parameters by Kinsella (1993)

Multi-Energy Method Parameter	Peak Overpressure bar(a)
10	>10
7	1.0
6	0.50
5	0.20
4	0.10
3	0.05
2	0.02
1	0.01

When considering the effects of blast waves on surrounding buildings, blast wave reflection factors ranging from 2 to 4.5 have to be used. Curves showing the reflection factors for different blast strength and angle of incidence can be found in Baker et al. (1983, p. 87) and Glasstone (1962).

Applying the multi-energy and Baker–Strehlow methods consist of the following steps (Baker et al., 1997):

1. Systematically identify the areas of confinement and congestion in which a vapor cloud explosion could occur. Call these potential explosion sites, PESs. Consolidate adjacent and closely separated PES areas. Calculate area of each.
2. Use a dispersion model to plot the plan view footprint contours of the LFL and UFL. Lay out this footprint on the plot plan beginning at various source points and using various wind directions and wind speed, atmospheric stability classes.
3. Estimate cloud height and/or confinement height over each PES.
4. Calculate flammable mass in each PES using the volume of the PES and the fraction of the cloud that is within the PES (see Chapter 5).
5. Calculate the average heat of combustion in the flammable portions of the cloud.
6. For the Baker–Strehlow method, select the reactivity of the fuel (low, medium, high). See Section 4.8.5. (This is not used explicitly with the multi-energy method.)
7. Select the level of confinement for each PES (1D, 2D, 3D). See Section 4.8.6.
8. Select the level of congestion for each PES (low, medium, high). See Section 4.8.7.
9. Find the parameter on the blast prediction curves. See Section 4.8.8.
10. Calculate explosion characteristics and predict blast loads on buildings. Superimpose separate blast waves when there are multiple PESs and multiple ignition sources within a single vapor cloud.

Since the first four steps involve calculating the flammable mass in the cloud, they are discussed in Section 5.3. Definitions for Steps 6 through 8 are given below. Example cases are developed in Chapter 7.

4.8.5. Determining Fuel Reactivity

Step 6 in using volume-source explosion models is to select the reactivity of the material. The reactivity ratings defined by Zeeuwen and Wiekema (1978) were adopted into the Baker–Strehlow method. These use the following criteria, based on the laminar burning velocity:

Low reactivity	$S_u < 45$ cm/s
Medium reactivity	$45 < S_u < 75$ cm/s
High reactivity	$S_u > 75$ cm/s

This definition is reflected in the grouping of materials listed in Appendix C.

To estimate the laminar burning velocity for mixtures, Baker et al. (1997) suggest using the LeChatelier equation in terms of the mole fraction of each flammable vapor component, y_i, and the laminar burning velocity of each component, S_{ui}, in cm/s:

$$S_u = \frac{\sum_i y_i}{\sum_i (y_i / S_{ui})} \tag{4.79}$$

This equation is approximate and the user should be conservative in using it (round upward and accept values close to the defined threshold as the higher reactivity).

4.8.6. Determining Degree of Confinement

As defined in Section 4.8.1, confinement, as used here, refers to the dimensions available for flame expansion. Areas with 1D or 2D confinement are considered PESs. Sources of confinement that create 2D flame expansion include solid decks, roofs, fin-fan coolers, and extremely densely packed pipe racks (>80% blockage). Often these structures are quite high relative to cloud heights for flammable heavy-gas clouds. At issue is whether during combustion, the cloud will expand sufficiently to drive unburned flammable mixture up to the height of 2D confinement where the confinement will contribute to flame acceleration. For common hydrocarbons the expansion ratio of combustion products to unburned fuel is a factor of 7–8. An expansion ratio of 8 represents an increase in radius of a hemispherical or spherical cloud by a factor of 2. Therefore, a source of 2D confinement that is below 2 times the height of a vapor cloud will influence the combustion event. Thus, as a rule of thumb, use 2D confinement when the confinement source is less than 2 times the cloud height.

Another effect of confining surfaces is reflection of shock waves. Using a 2-dimensional model of blast waves, Baker et al. (1994) concluded that shock wave reflection from a solid plate at least 20 m above the ground has no effect on either overpressure or blast impulse at locations on the ground in either the near-field or the far-field. On the other hand, a plate at 10 m above the ground introduces a reflection that increases overpressure at all distances. Impulse strength is increased in near the plate. Since the 20 m height represents 3.2 cloud radii for the test case used, the recommended rule of thumb is that confinement above 3.2 times the cloud height can be treated as inconsequential to the vapor cloud explosion.

Step 7 in Section 4.8.4 requires selecting a single degree of confinement that represents the entire area when there are varying levels of confinement in the area. This selection proved to be particularly troublesome when the selection is between 2D and 3D confinement. This is seen in Table 4.10, where, for example, with fuel of medium reactivity, and with medium obstacle density, the flame speed parameter values for 2D and 3D expansion are 0.662 and 0.10. The corresponding predicted overpressure is 1.7 and 0.06 atm, a substantial difference. To resolve this problem, Baker et al. (1994) suggest interpolating between 2D and 3D to create an intermediate 2 ½ D level of confinement. For the above example, the average flame speed parameter value for 2½ D would be Mach 0.381

4.8.7. Determining Level of Congestion

The degree of congestion is the parameter most difficult to quantify. The volume blockage ratio (VBR), the ratio of the volume occupied by obstacles to the total volume, was introduced in Section 4.8.1. Other experimenters have used the area blockage ratio (ABR) defined as the area blocked by obstacles to the total cross-sectional area, and the pitch, which is the distance between successive obstacles or obstacle rows. There is an optimum pitch for flame acceleration. When the pitch is too large, the wrinkles in the flame front will burn out before the next obstacle is reached. When the pitch is too small, the gas pockets between successive obstacles are relatively unaffected by the flow. Consequently, it is conservative to consider obstacles only with the optimum pitch. This leaves ABR and VBR as the variables of choice.

Unfortunately, most congested areas in refineries and chemical plants have many types of obstacles including pumps, structural members, pipes, valves, vessels, and electrical equipment. The spacing between these types of obstacles is very irregular. A pipe rack is one of the few structures in which a regular pattern is found, allowing direct application of experimental data.

The following definitions of degree of congestion are suggested by Baker et al. (1994) for Step 8 of the explosion modeling process (Section 4.8.4).

Low congestion:	ABR <10%, obstacles widely spaced, <3 layers of obstacles
High congestion:	ABR > 40%, obstacles fairly closely spaced, ≥3 layers of obstacles
Medium congestion:	Between low and high

Twelve European companies sponsored a project to find better correlation parameters between degree of congestion and the maximum explosion overpressures. Experimental projects MERGE and EMERGE were designed to obtain a comprehensive and consistent database on the influence of obstacles on explosion peak overpressure (Mercx et al., 1994). A companion project, called GAME (guidance for the application of the multi-energy model)(van den Berg and Eggen, 1996) devised a correlation for peak overpressure as a function of measures of congestion. The correlation is (Eggen 1995; Mercx, 1998):

$$\frac{\Delta P}{P_0} = a\left(\frac{\mathrm{VBR} \cdot L_\mathrm{f}}{D}\right)^b S_\mathrm{u}^{2.7} D^{0.7} \tag{4.80}$$

where

VBR = volume blockage ratio (–);
D = average obstacle size (m);
L_f = maxiumum flame path length (m);
S_u = laminar burning velocity (m/s);
b = 2.75 for 3D, 2.25 for 2D confinement
a = 0.84 for 3D, 3.38 for 2D confinement

The term L_f is the distance a flame can travel and therefore accelerate within the congested area. In a worst-case, if ignition occurs at a boundary to the congested area, it is the length of the congested area. Since the tests were made

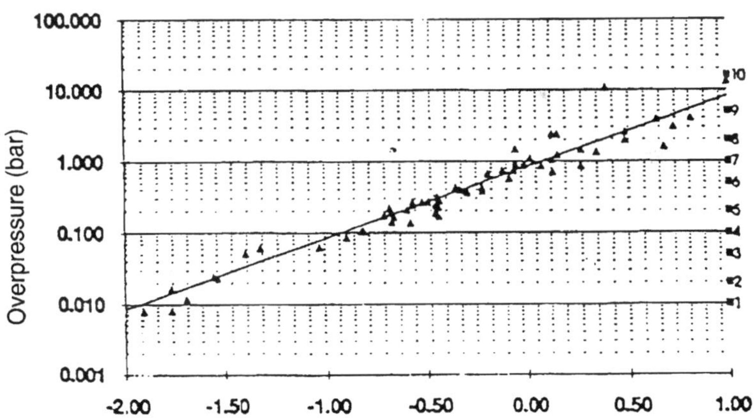

$$\left(\frac{\mathrm{VBR} \cdot L_\mathrm{f}}{D}\right)^{2.75} S_\mathrm{u}^{2.7} D^{0.7}$$

Figure 4.59. Peak overpressures measured in project MERGE tests versus the project GAME fuel reactivity and congestion parameter for 3-dimensional expansion tests. (Mercx and van den Berg, 1997.)

with central ignition, it is more appropriate to take L_f as ½ the maximum length in the congested area. The size of the obstacle arrays used in the MERGE experiments is 2, 4.5, and 9 m, so L_f would be 1, 2.25, and 4.5 m. The values of D used in the MERGE experiments are $D = 0.019, 0.041, 0.082, 0.168$ m. The VBR values used are VBR = 0.05, 0.10, 0.20.

Data from the MERGE project are shown in Figure 4.59 plotted in the form of the Project GAME correlation, Eq. (4.80). Equation (4.80) constitutes a tentative extension of the multi-energy method which is being evaluated in a follow up project GAMES (GAME Second phase).

4.8.8. Multiple Congested Volumes

Adjoining congested areas should be consolidated into one. Areas separated by a small enough distance should also be consolidated. Figure 4.52 discussed in Section 4.8.1 shows that flames decelerate and then reaccelerate in passing through a 2 m gap free of obstacles. For a gap of 3 m or more, flame speeds decrease enough to produce separate explosion peaks, so we recommend consolidating congested or confined areas separated by less than 3 m. Other researchers advocate using a larger separation threshold for consolidating PESs. Baker et al. (1997) recommends 5 m (16 ft.).

Should the flammable mass within 3 m of a PES be added to the flammable mass within the PES for calculating explosion effects? Most knowledgeable practioners say no. The argument is that during the explosion the expanding burned gases tend to push flammable gases out of the PES, so some of flammable gases are burned in an uncongested/unconfined area. Thus, it is overly stringent to include even all of the flammable gases in the PES, let alone gases outside the PES. Again, other researchers use larger separation thresholds. The *Yellow Book* (1997) description of the multi-energy method says that if the distance between two obstructed regions is 25 m (82 ft.) or more, the combustion energy in the unobstructed region can be neglected for determining explosion energy.

A large vapor cloud could encompass several PESs. The multi-energy method predicts that blast waves will originate from the individual PESs, but not in open areas. Because of the variability in ignition location, wind direction, etc., several overlapping pressure-time history combinations can be produced. For example, Figure 4.60 (from Baker et al., 1997) shows a vapor cloud encompassing three alternative ignition sites, A, B, and C. Ignition will generate two explosions centered in areas of high and low congestion. The pressure curves experienced by the observer in the lower left hand corner are illustrated in Figure 4.61. The more damaging case is B, in which the overpressure curves tend to overlap.

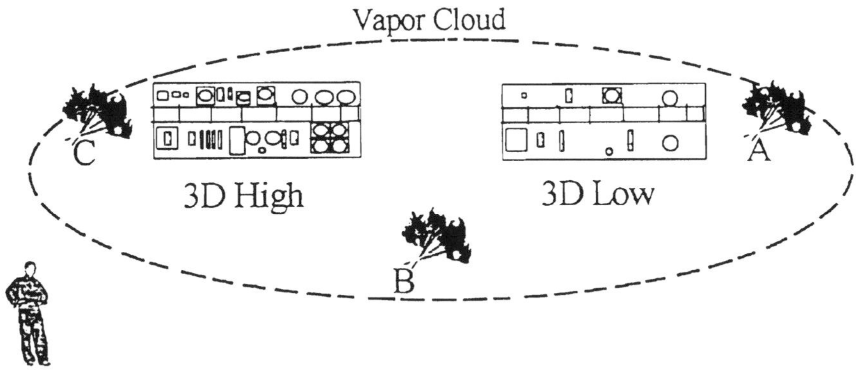

Figure 4.60. Schematic example of ignition of a vapor cloud (Baker et al., 1997).

4.9. Minimum Flammable Mass for Vapor Cloud Explosions

The question arises, is there a minimum flammable mass needed for a vapor cloud explosion to occur? The question requires further qualification, since it is common experience that, for instance, a few drops of flammable vapor can explode in a laboratory shock tube. Rephrasing, then, is there a minimum vapor cloud size for an outdoor release that has a nonnegligible probability of ignition and flame acceleration to deflagration flame speeds in plant geometries allowing two-dimensional and three-dimensional flame expansion?

Some older rules of thumb set fairly high values for a minimum flammable mass capable of producing damage. However, we show in examples fairly low values, given sufficient confinement or congestion. Clearly, the minimum flammable mass must be defined relative to local geometries.

There are reasons to expect a minimum to the flammable mass in a vapor cloud capable of exploding:

- Since ignition sources are distributed over areas, if the density of ignition sources is constant, small vapor clouds have a lower probability of encountering an ignition source.
- Small explosions are less damaging than large ones. Given that pipes and equipment in plants have a certain minimum strength, there is a minimum flammable mass likely to cause explosion damage.
- Since turbulence-inducing structure is needed for flame acceleration to explosions (as discussed in Section 4.8.1) there is likely to be a minimum flammable volume consistent with the size and spacing of typical structural elements.

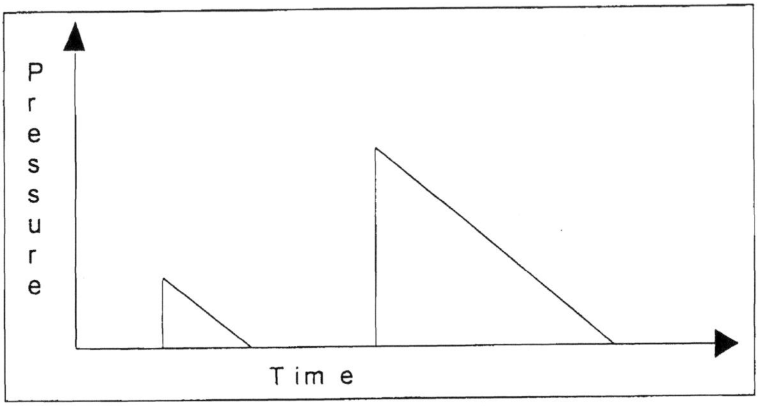

Pressure Curves for Ignition at Point A

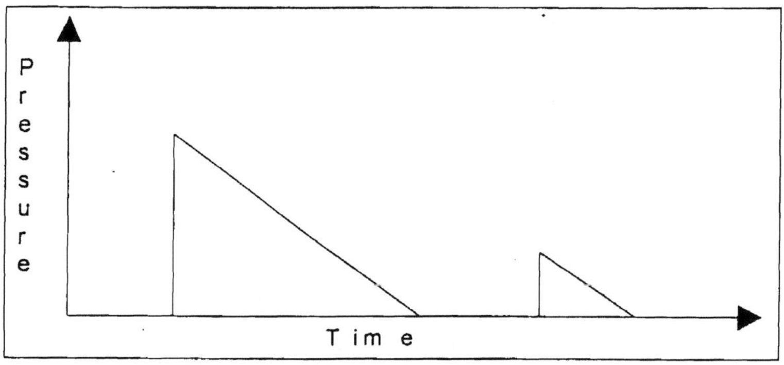

Pressure Curves for Ignition at Point C

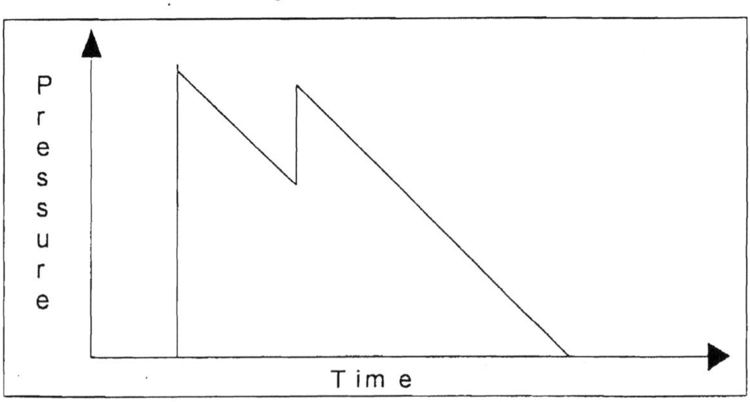

Pressure Curve for Ignition at Point B'

Figure 4.61. Example of explosion pressure curves reaching observer after ignition at points shown in Figure 4.60 (Baker et al., 1997).

In practice, to define a minimum flammable mass of concern, first specify local geometry and define a tolerable damage level. This could be, for example, 10% window breakage from an explosion centered in a nearby shed or pipe rack. Using a correlation or damage model, translate this criterion into a tolerable maximum overpressure and/or impulse level. Then model a realistic vapor cloud located in the specified congested/confined area. Use a dispersion model to calculate concentration gradients and flammable mass in vapor clouds of various sizes. To be realistic, locate the center of the explosion at some distance from the edge of the cloud, such as at the point of stoichiometric concentration or at the center of the flammable cloud. Calculate explosion characteristics with the methods shown earlier. Find the largest vapor cloud that will produce overpressures and impulses just short of the specified levels.

Alternately, a minimum flammable mass can be estimated from probability of ignition/explosion plots discussed in the next section.

Further experimental work is needed to realistically identify a minimum flammable mass as discussed in Chapter 8.

4.10. Probability of Vapor Cloud Ignition and Explosion

The probability of ignition decreases as the area of the flammable cloud decreases, simply because the probability of encountering an ignition sources decreases with exposed area. Another reason is that as the mean concentration of a vapor cloud decreases, the pockets of gas above the lower flammable limit become more and more isolated, with decreasing chance of a flame propagating from one to the next.

Indeed, historical experience bears out that that the probability of a vapor cloud exploding is a function of cloud size. Figure 4.62 (replotted from Prugh, 1987) plots the probability of explosion, p_E, vs. the flammable mass, f_m, from historical explosion data by Davenport (1977) and Kletz (1977). Prugh also cites Giesbrecht (1980), Lees (1980), and Lind (1975) for developing correlations of p_E with f_m.

The basis for Figure 4.62 is that a 100,000 kg release of flammable material has a probability of explosion of 50%. According to Figure 4.62, the probability of explosion drops below 0.1% when the flammable mass is below about 250 kg.

In a review of ignition probability models, Spencer and Rew (1997) found several models calculating the probability of vapor cloud ignition as a function of cloud area. This is illustrated in Figure 4.63 which shows the predictions of models by Simmons (1975), Cox et al. (1990), and the UK Health & Safety Executive (HSE) RISKAT model (Clay et al., 1988).

The probability of ignition reaches near zero levels in Figure 4.63 for cloud areas ranging from 24 m^2 (Simmons delayed ignition) to about 300 m^2 (HSE

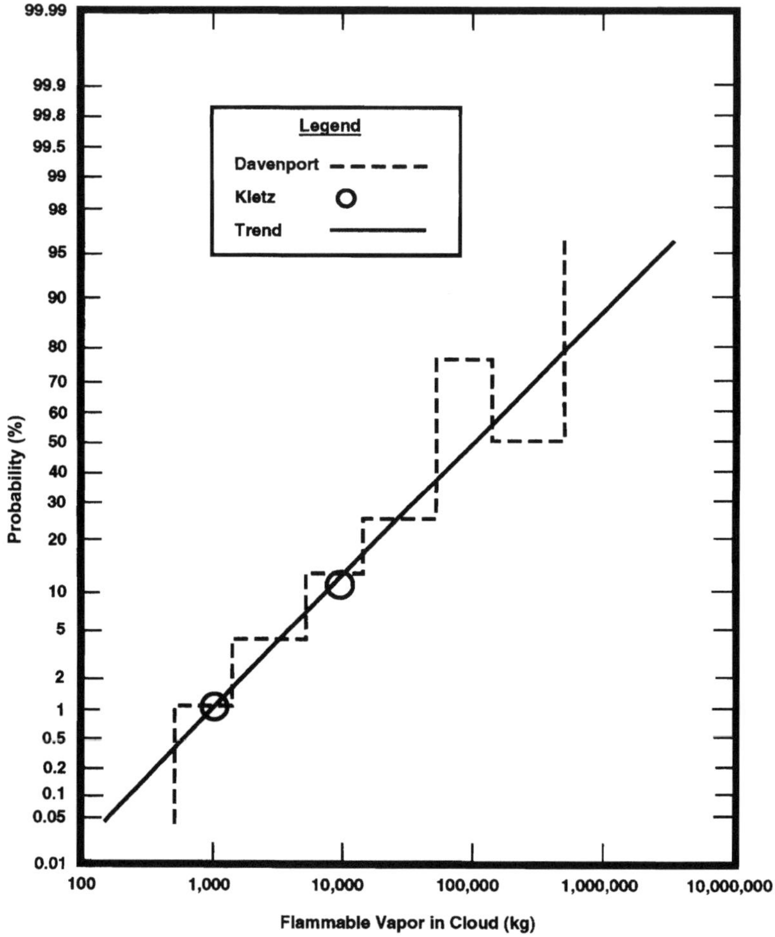

Figure 4.62. Probability of vapor cloud explosion shown as function of flammable mass in release.

industrial and urban areas) to around 6000 m^2 for the HSE model in rural areas. The Simmons model for immediate ignition is not pertinent for estimating a minimum flammable mass.

The main difference in the Simmons and HSE models is in their assumption of ignition source density. This is assumed constant in the HSE model at values shown in Figure 4.64. The ignition model of Simmons assumes that the ignition density varies with cloud area as shown in Figure 4.64.

The HSE RISKAT model makes use of an assumption similar to Prugh's, that a large instantaneous release of 200 tonnes of LPG which has drifted

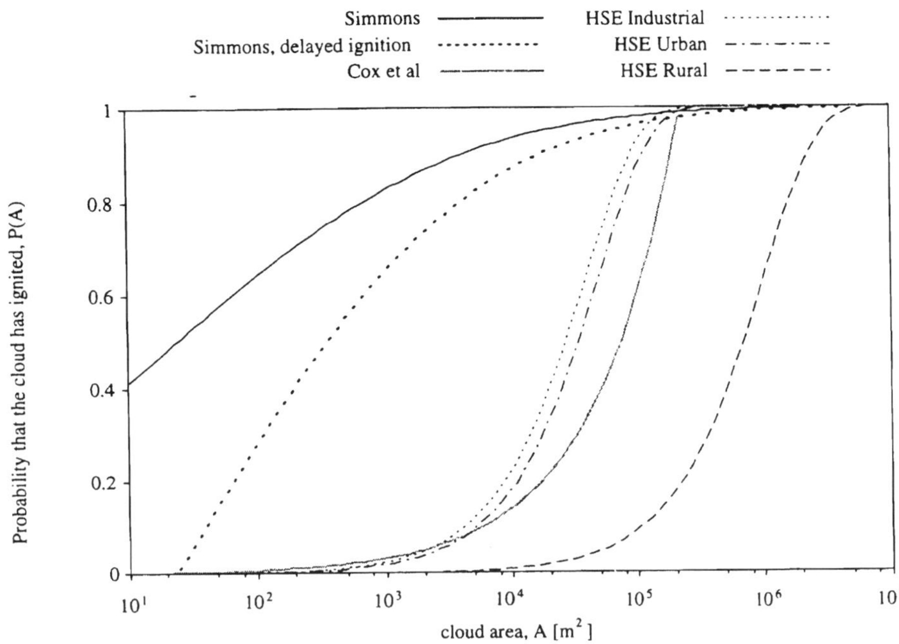

Figure 4.63. Comparison of several models of ignition probability as a function of vapor cloud size. (Spencer and Rew, 1997) (Reproduced by permission of Controller, HM Stationery Office.)

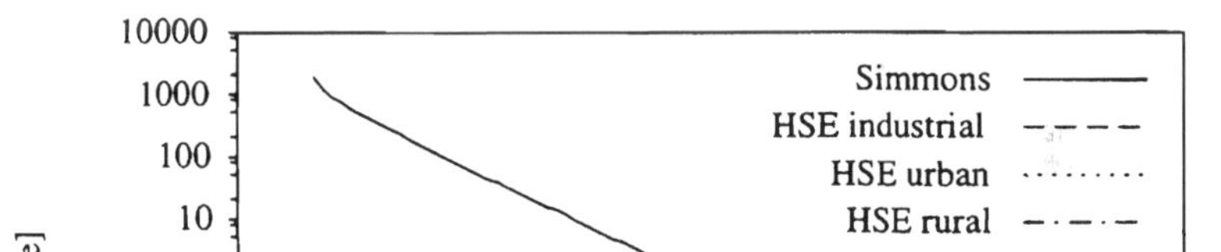

Figure 4.64. Variation of ignition source density with cloud area for the Simmons (1974) and HSE ignition models (Spencer and Rew, 1997). (Reproduced by permission of Controller, HM Stationery Office.)

downwind over industrial plants in D stability, 5 m/s wind speed weather has a probability of nonignition of $Q_f = 1 \times 10^{-6}$. The probability of ignition is assumed to be a function of the ratio of the cloud area at any point in time, A, to the area swept by the flammable cloud if it disperses to its full size, A_f. The probability of ignition is given by

$$P(A) = 1 - (1 - P_f)^{A/A_f} \tag{4.81}$$

This equation is simpler expressed as the probability of nonignition, Q, where $Q = 1 - P$, as

$$Q(A) = Q_f^{A/A_f} \tag{4.82}$$

This can be written in terms of the average ignition source density, μ, as

$$Q(A) = e^{-\mu A} \tag{4.83}$$

Inverting this equation gives μ in terms of the base area, A_f, and base nonignition probability, Q_f:

$$\mu = -\frac{1}{A_f} \ln Q_f \tag{4.84}$$

A disadvantage of the HSE model is that it is sensitive to the assumption of the value of Q_f, which is supposed to be a negligible level, but it could range from 1×10^{-4} to 1×10^{-8}.

For the Simmons relationship μ is linear in A, Q is proportional to the $\exp(-k * A^2)$ where k is about 1.5, so the range is between a near zero probability of ignition and near 100% probability.

5

DETERMINATION OF FLAMMABLE MASS

Every dog is allowed one bite, but in industry we are expected to muzzle our dog if a similar dog has bitten someone elsewhere.
—Kletz, 1990

This chapter provides useable formulas to address Step 4 in the flowsheet in Figure 2.1 (calculate flammable mass in vapor clouds). Methods detailed in Chapter 4 address Step 3 in the flowsheet showing how to calculate the total mass of flammable material released as a function of time. The next problem is to determine the flammable mass in a vapor cloud by integrating across the concentration profiles between two concentration limits, the LFL and the UFL. Actually, this can be broken into two separate steps, depending on whether the postulated scenario is a flash fire or an explosion:

1. For a flash fire, calculate the mass of flammable material that is above the LFL.
2. For explosions, in addition find and subtract the mass of flammable material that is above the UFL.

In flash fires, the rich portion of the cloud will eventually burn as a diffusion flame after the premixed portion has burned. So, for flash fires the flammable mass is all the mass above the LFL. However, explosions occur so rapidly that the mass of flammable material in the rich portion of the cloud does not contribute to the explosion, except as the flammable limit boundaries are dynamically changed during the explosion by mixing, pressure, and temperature effects. Thus the flammable mass for explosions is approximately the mass of fuel between the LFL and the UFL.

5.1. Estimation Methods by Degree of Confinement

Various models differ in their cloud concentration profiles by using different boundary conditions and assumptions. The choice of which model to use depends on the degree of confinement, primarily on whether the release is indoors or outdoors. A logic diagram in Figure 5.1 leads to one of the following three sections:

- Unconfined clouds, Section 5.2
- Partially confined clouds, Section 5.3
- Confined releases, Section 5.4

Each of these sections provides several modeling choices. The simplest methods are usually preferred for screening studies and the more complex methods are preferred for more accuracy. The simpler methods are treated first.

5.2. Methods for Finding the Flammable Mass in Unconfined Vapor Clouds

Unconfined vapor clouds are those for which solid boundaries, excluding the ground, and structural elements do not appreciably influence the cloud expansion. There can be some structural elements present. If they are present in

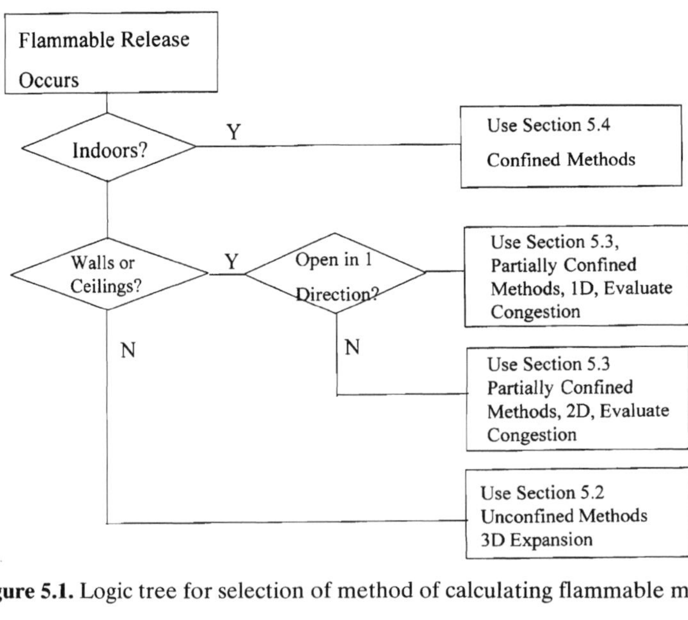

Figure 5.1. Logic tree for selection of method of calculating flammable mass.

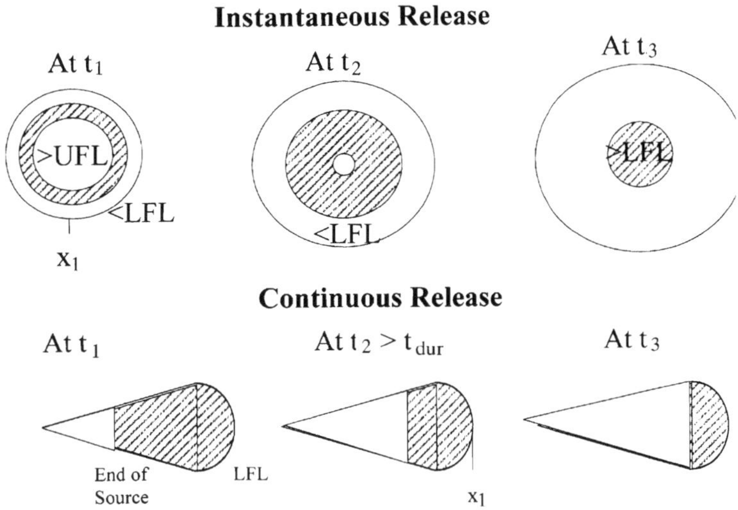

Figure 5.2. Sketch of flammable mass region of vapor clouds as cloud moves downwind.

sufficient density, they can contribute to the degree of congestion and to turbulence that increases dispersion and flame speeds.

Figure 5.2 illustrates for instantaneous, continuous, and finite-duration releases how the flammable portion of the cloud changes with time. Consider mean concentrations and not fluctuating values. For an instantaneous release, if the cloud is released without considerable initial mixing, then all the concentrations in the cloud except near the edges will be initially above the UFL. The flammable region will be a relatively thin shell around the surface of the cloud with its inside contour at the UFL and its outside contour at the LFL. As the cloud travels downwind and dilutes, the shell of flammable vapors expands and the mass in the flammable region increases. Just as the shell reaches the center of the cloud the amount of vapor or aerosol in the flammable region will reach a maximum. With further dilution, the flammable mass decreases. When the outer, LFL contour reaches the center of the cloud the flammable mass becomes zero.

For a continuous steady release of flammable material continuing even 5 minutes, the duration is considered long enough to develop a steady-state plume, uninfluenced by release duration. If the release duration is less than about 5 minutes, it is considered a finite-duration release. The resulting plume is a truncated steady-state profile after the source stops. The front nose of the flammable cloud (to the LFL) remains essentially stationary as the back end of the flammable region moves forward. When the back end reaches the front end the cloud has diluted below the LFL. The release duration has a strong effect on the plume dispersion.

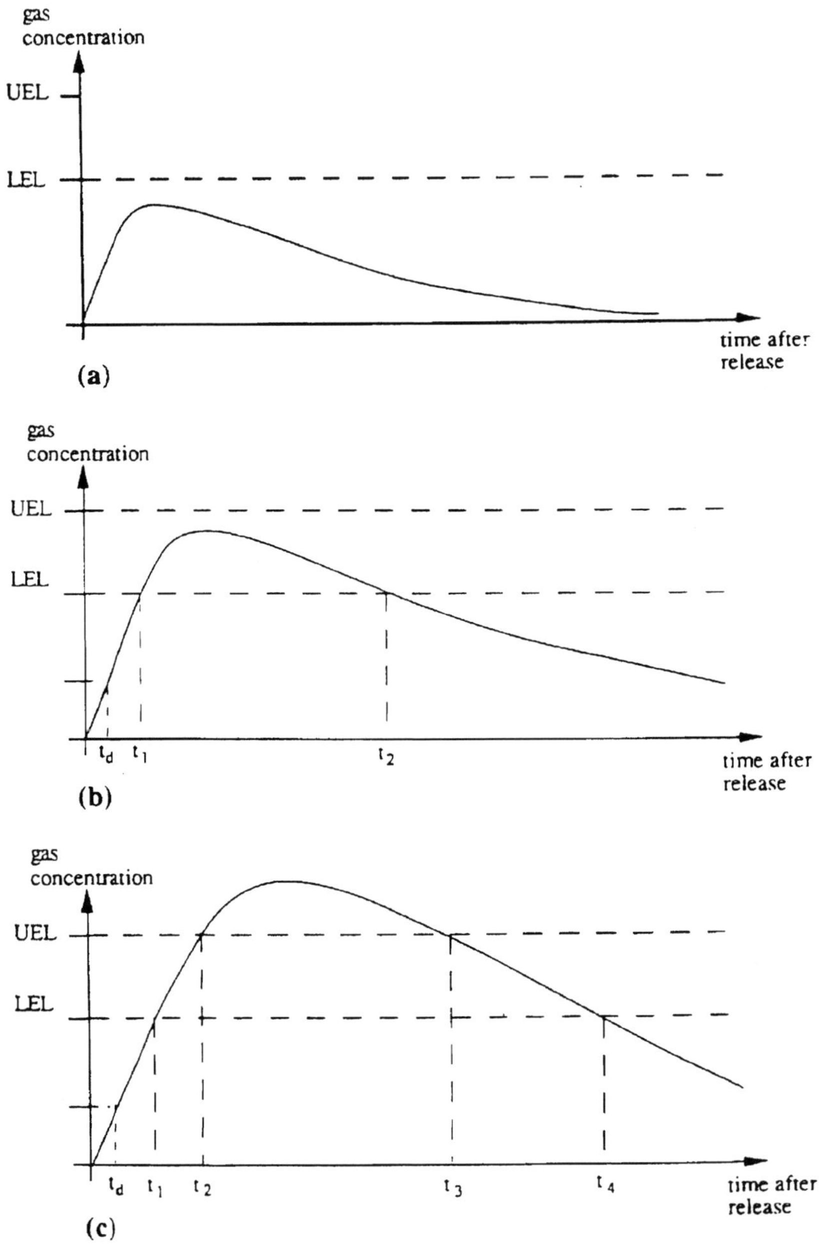

Figure 5.3. Typical concentration time histories for a continuous release: (a) The concentration always lies below the LFL; (b) The concentration exceeds the LFL but not the UFL; (c) The concentration exceeds the UFL. (Reproduced by permission of IChemE.)

Figure 5.3 illustrates concentrations at a given distance downwind from a time-dependent discharge rate. Since discharge rates typically decrease with time, and since the front nose of a cloud has more diffusion than the later part of the cloud, the concentration increases at first and then decreases. The peak concentration can be below the LFL, above the UFL, or in between.

Methods of estimating flammable mass for unconfined vapor clouds follow below.

5.2.1. Screening: Rules of Thumb

There are no simple formulas to estimate the fraction of mass that is in the flammable range for an accidental release of material. Flammable mass is treated by various government agencies only by reference to the inventory of flammable material in a plant, as summarized below.

EPA RMP Guidelines

The U.S. Environmental Protection Agency (EPA) issued requirements (EPA, 1996a) for development of risk management programs and plans (RMP) at facilities that handle more than a threshold quantity of certain toxic or flammable substances in a process. One guideline is generic guideline and one specific to anhydrous ammonia (EPA, 1996b). They complement requirements promulgated by the Occupational Safety and Health Administration (OSHA) on process safety management (PSM). Under the RMP rule there are 77 toxic substances and 62 listed flammable substances (gaseous and highly volatile). The flammable substances all have a threshold quantity of 10,000 pounds (4536 kg). In other words, if one has more than 10,000 pounds of a listed flammable substance (e.g., propane, butane) in a process, one must develop a risk management program and plan. One element of this program is an assessment of the consequences of a worst case release scenario and a more realistic alternative release scenario, documented in the RMP. The worst case release is designed to represent the upper bound catastrophic, conservative situation; it assumes complete release of the quantity held in the largest vessel at the facility and assumes (contrary to the laws of physics) that the entire mass is available for a vapor cloud explosion. The alternative release scenario is designed to represent a more likely emergency release situation. In this case, smaller release quantities are expected and flammable mass calculations can be applied.

EPA has developed and published guidance for the analyses of offsite consequences associated with the worst case and alternative scenarios. The Offsite Consequence Analysis (EPA, 1998) provides a variety of simple calculation tools for flammable release scenario modeling such as TNT equivalent and multicomponent methodologies. The purpose of this guidance is to make available easy-to-use models for a wide variety of businesses who may not have sophisticated processes or personnel capable of handling complex analytical tools. Although these models are simple, they have value for screening and focussing in on critical areas where more sophisticated analytical techniques should be used. Use of these tools is not mandatory; any appropriate model may be used for these assessments as long as certain input parameters and requirements are met.

Dutch Government Guidelines

The government of the Netherlands defined screening-level threshold quantitites of both flammable and toxic materials to determine when a risk assessment is required for new plants or expansions. Their potential-hazard indexing system is based on threshold quantities and correction factors. Correction factors provide credits for storage in enclosure, for storage temperatures below the normal boiling point, and for process temperatures below 25 C (68 F). They provide a penalty for storage above the boiling point. The threshold quantity of a hazardous substance is defined as that quantity which when released suddenly under reference conditions can result in extreme danger for a worker present within a radius of 100 m from the point of emission.

- For flammables the threshold quantity is 10,000 kg
- For extremely toxic substances the threshold quantity is 1 kg
- For toxic substances the threshold quantity is 300 kg for the reference compound, chlorine.

Again, these guidelines do not attempt to calculate actual flammable mass, but merely refer to flammable inventory compared with a threshold of concern.

EEC Occupational Safety Report

The Seveso Directive (European legislation) designates installations for which a safety report is mandatory. The EEC requirements are similar to the Dutch system, referencing flammable mass inventories and thresholds of concern.

5.2.2. Calculating FLammable Mass with Dispersion Models

Dispersion models generate predictions for vapor cloud dimensions and concentration profiles which are admittedly over simplified but nonetheless mathematically precise. Taking advantage of this mathematical precision the predicted flammable mass in a vapor cloud can be found by integrating over the cloud volume between the flammable limits. This integration is not a trivial exercise, so a computer program is needed for all but the simplest "top hat" dispersion models.

As discussed in Chapter 4, there are three dispersion mechanisms:

- jet mixing (applies very near source with high initial velocity)
- heavy gas dispersion (applies after jet mixing decays while cloud is heavier than air)
- passive or Gaussian dispersion (applies after cloud is dilute and neutrally buoyant)

The heavy gas dispersion regime does not occur with lighter-than-air gases, hydrogen, methane, synthesis gas, etc., The preferable models to treat the dispersion of a flammable cloud are models that integrate jet dispersion and heavy gas dispersion mechanisms. These models usually also treat passive dispersion, but a flammable cloud has usually diluted below the LFL by the time the passive mode becomes the dominant mechanism. A few of the preferred types of models are discussed here to illustrate the concentration profiles they predict. In each case, flammable mass is found by numerical integration across the concentration profiles. The following are treated in subsequent subsections:

- Top hat models—flat concentration profiles
- Three heavy-gas dispersion models which use numerical integration of flammable mass. See Appendix L for details of numerical integration to find flammable mass.
- A method of modeling successive Gaussian puffs

Although passive dispersion models do not adequately represent the concentration profiles and dispersion mechanisms typically dominant in flammable vapor clouds, nonetheless, for passive models analytic solutions are available for calculating flammable mass. Even so, these solutions are so complicated as to require a computer program. The solutions are given for the following cases:

- Instantaneous release (Appendix I)
- Continuous release using approximate integration method (Appendix J)
- Continuous release using rigorous integration method (Appendix K)

Flammable Mass from Top Hat Dispersion Models

The earliest and simplest heavy gas dispersion models, describing both gases and aerosols, are the top hat type. This is named after their assumption that an instantaneous release assumes a cylindrical form. Concentration profiles (spatial variations) within the cloud are ignored; the concentration profile is assumed to be constant anywhere inside the cloud and zero outside but varies with time. For a continuous release the top hat profile is wedge-shaped. For a finite duration release the cloud is a truncated wedge. These are sketched in Figure 5.4. The cloud height is shown to slump as a heavy gas, and then increase as the cloud grows by entrainment.

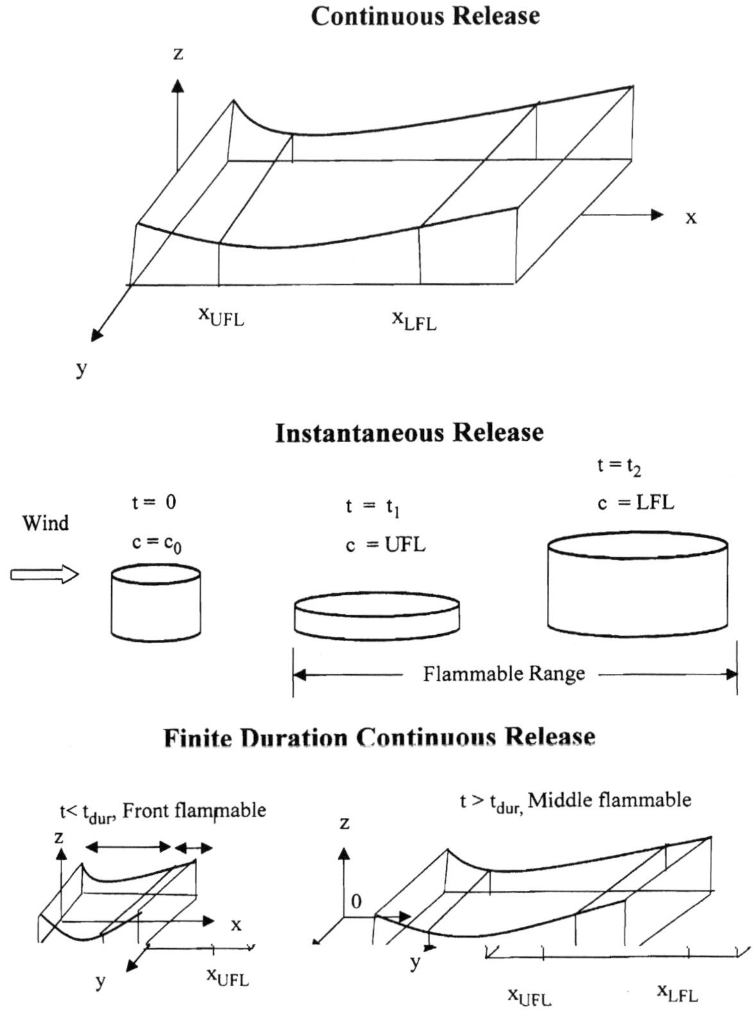

Figure 5.4. Illustration of top hat model forms and corresponding flammable volume.

The top hat model assumes there is no concentration gradient at the cloud edges, albeit air entrainment occurs through the top and edges. Thus, the flammable mass for an instantaneous release is always simply the total mass released as long as the cloud is in the flammable range. As the cloud drifts downwind it dilutes until it first reaches the UFL. At that time and distance it becomes flammable, and stays flammable until the cloud is diluted below the LFL.

For the general case of a finite-duration continuous release with a constant release rate, w (kg/s) the cloud develops a concentration profile that decreases from the back end of the cloud to the front. This concentration profile remains essentially stationary but material in any given vertical plane moves from the back to the front. When the discharge stops, the flammable back end of the cloud moves along until it reaches the front end of the cloud, and after that the cloud concentrations all along the cloud dilute below the LFL. The flammable mass in the cloud is simply the total mass released in the appropriate interval between when the cloud profile dilutes below the UFL and yet is above the LFL. The models indicate the distance at which the average concentration reaches these limits, say at x_{UFL} and x_{LFL}. The corresponding times at which the cloud front reaches these distances are t_{UFL} and t_{LFL}. These are found in general by integrating the time-variable cloud speed, but for the simple case of a constant cloud speed the flammable mass, m_{F}, is given by

Finite-duration, continuous release, top hat model

$$m_{\mathrm{f}} = \begin{cases} w(t - t_{\mathrm{UFL}}) & x > x_{\mathrm{UFL}} \\ w(t_{\mathrm{LFL}} - t_{\mathrm{UFL}}) & x_{\mathrm{LFL}} > x > x_{\mathrm{UFL}} \end{cases} \tag{5.1}$$

$$x_{\mathrm{UFL}} = u_{\mathrm{cld}} t_{\mathrm{UFL}} \tag{5.2a}$$

$$x_{\mathrm{LFL}} = u_{\mathrm{cld}} t_{\mathrm{LFL}} \tag{5.2b}$$

Top hat models are useful, conservative first approximations. Since the only difficult task for a top hat model is to calculate air entrainment rate, and since this task requires very few tuning parameters, top hat models can usually do this task quite well. Since they do not account for dilution at the edges which takes part of the released mass outside of the flammable range, they tend to overestimate the flammable mass. In the near field when the centerline concentration is high, the overestimation error is quite small, typically $<10\%$. In the far field when the centerline concentration is near to the LFL, the overestimation error of top hat models can be quite high, $>70\%$. Other models that represent a more general concentration profile, such as Figure 5.6, discussed later, provide a basis for estimating top hat modeling errors.

Heavy Gas Dispersion Effects

The behavior of heavy gas clouds differs significantly from that of passive clouds in two important respects:

1. Heavy clouds can spread laterally, downwind, and even upwind against a light wind by their negative buoyancy, much like a liquid spill.
2. Air entrainment rates are reduced by stable density stratification within the cloud.

As pointed out by Spicer and Havens (1996), the range in which heavy gas effects are significant can be indicated for low momentum releases (discharge velocity about equal to or less than wind speed) by the release Richardson number:

$$Ri_c = \frac{g}{u_*^2}\left(\frac{\rho_0 - \rho_{air}}{\rho_a}\right)\frac{w}{\rho_0 u_w D} \tag{5.3}$$

where u_* is the friction velocity, ρ_0 is the initial density of the discharged material, ρ_{air} is the air density, w is the discharge rate, u_w is the wind speed, and D is the diameter of the source. When $Ri_c > 1$ the heavy gas effects are important and a passive dispersion model should not be used. This is illustrated later in the section on flash fire ignition tests.

Heavy Gas Dispersion Models

There are a number of models that treat both heavy gas and passive dispersion, some of which are listed in Chapter 6. Here we discuss briefly the aspects of representative models salient to their estimation of flammable mass, namely how they calculate concentration profiles and how they integrate these profiles between the flammable limits. The models treated are PHAST (UDM), DRIFT, DEGADIS, HEGADAS, and FEM3.

PHAST and DRIFT Models

The models PHAST (**p**rocess **ha**zard **s**oftware **t**ool, DNV Technica, 1996) using the unified dispersion model (UDM) modification, and DRIFT (Webber et al., 1992) use similar methods. They model either ground level or elevated releases. For an elevated, heavy vapor/aerosol release the cloud is modeled initially as either a sphere (instantaneous release) or a tube (continuous release) with a circular cross section that tends to flatten into an ellipse as the cloud settles. Upon touching the ground. The cross section becomes a truncated ellipse, and the vertical component of momentum is dissipated. Aerosol droplets may rain out shortly after touchdown. Rainout produces a pool which spreads and vaporizes. If spilled onto water, part of the material may also dissolve. The vapor from the evaporation of the pool is added back to the plume as a function of time. The plume can become buoyant after evaporating or raining out all aerosol droplets and picking up heat by ground conduction, or by condensing water picked up over a wet surface. A buoyant plume lifts off and rises until constrained by the mixing layer. The mixing layer is an effective ceiling for atmospheric dispersion caused by an abrupt change in the vertical temperature gradient in the atmosphere. It acts to trap pollutants below the mixing layer.

PHAST and DRIFT use generalized profiles that have sharp-edged concentration profiles near the source, gradually diffusing to Gaussian profiles in the far field. They assume the concentration profiles of the form given by

Continuous (or Finite Duration) Source:

$$c(x, y, z) = c_0(x)F_z(x, z)F_y(x, y) \tag{5.4a}$$

$$F_z(z(x)) = \exp\left[-\left(\frac{z}{R(x)}\right)^s\right] \tag{5.4b}$$

$$F_y(y(x)) = \exp\left[-\left(\frac{y}{W(x)}\right)^m\right] \tag{5.4c}$$

Here $c_0(x)$ is the centerline concentration at any value of downwind distance, x. The scaling is provided by $c_0(x)$ and the lateral and vertical distribution of concentration is provided by the terms $F_z(z, x)$, $F_y(y, x)$, and $F_x(x)$ as a function of distance.

Instantaneous (Puff) Source

$$c(x, y, z) = c_0(x)F_z(x, z)F_y(x, y)F_x(x) \tag{5.5a}$$

The $F_z(z, x)$, $F_y(y, x)$ are as before and $F_x(x)$ is given in terms of the centerline position, x_0 as

$$F_x(x) = \exp\left[-\left(\frac{x - x_0}{W(x)}\right)^m\right] \tag{5.5b}$$

The scaling factors W and R are the characteristic half-width and a half-height, which are proportional to σ_y and σ_z, the cross-wind and vertical standard deviations.

$$W = \sqrt{2}\sigma_y \tag{5.6a}$$

$$R = \sqrt{2}\sigma_z \tag{5.6b}$$

Equations (5.4) and (5.5) reduce to the Gaussian form [Eqs. (I.2) to (I.5)] when $s = m = 2$. For larger values, say $m = 50$, the profiles predicted by Eqs. (5.4) and (5.5) are very nearly sharp-edged as Figure 5.5 illustrates. This formulation allows modeling of a sharp-edged jet, as occurs from a smooth-edged nozzle, dispersing to a jet with a more nearly Gaussian profile farther downwind.

The exponent m is correlated as a function of the normalized density difference $(\rho_{cld} - \rho_a)/\rho_a$. The correlation for s is similar to the correlations for atmospheric flux gradients proposed by Businger et al. (1971), or Dyer and Hicks (1970) as discussed by Lo and McBean (1978).

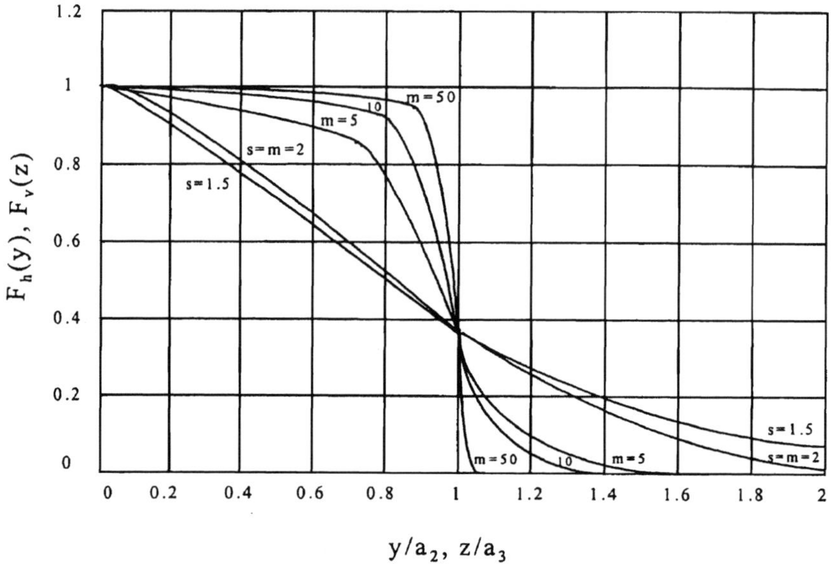

Figure 5.5. Vertical and horizontal concentration profiles for PHAST and DRIFT dispersion models.

The PHAST and DRIFT models use numerical integration of a set of simultaneous *ordinary* differential equations to predict air entrainment rate, vertical momentum, horizontal momentum, enthalpy, water evaporation from a substrate, and plume position. At each distance downwind, x, all of the parameters needed to define the concentration profiles given by Eqs. (5.4) and (5.5) are known.

The flammable mass is found by numerical integration of the concentration profiles between the flammable limits as discussed in Appendix L. Integration of Eq. (5.4c) is done with respect to y and of Eq. (5.4b) with respect to z at each downwind distance x. The numerical integration method successively cuts the integration step size until the resulting integrand changes very little with step size. The method is general and any concentration profile can be substituted for Eqs. (5.4b) and (5.4c).

DEGADIS and HEGADAS Models

The DEGADIS (**de**nse **gas dis**persion) model of Spicer and Havens (1989) and the HEGADAS (**hea**vy **gas dis**perion) model (Witlox et al., 1990) use a concentration profile in the cross-wind, y direction that is flat in the middle with Gaussian edges. They assume also a Gaussian profile in the vertical direction. A set of simultaneous ordinary differential equations is integrated to provide air entrainment rate, vertical and horizontal momentum, enthalpy, and cloud position. Flammable mass is found by numerically integrating across the concentration profiles between the flammable limits just as described for the PHAST and DRIFT models.

FEM3 Finite Element Model

The FEM3 (**f**inite **e**lement **m**ethod **3**) model developed by Lawrence Livermore National Laboratory (Chan, 1988; Gresho and Chan, 1984) numerically approximates the solution of the governing partial differential equations. It requires first establishing a numerical grid over the area of interest. The solution is found by numerical integration via the finite element method which requires considerable computing time. FEM3 is capable of predicting turbulence profiles as well as mean concentration profiles. It does this with an advanced k-ε turbulence closure model used routinely with finite-element models. The k-ε model uses two additional differential equations for turbulent energy generation, k, and decay, ε.

The predicted concentration profiles are discontinuous in that each finite element contains a different concentration prediction. Flammable mass is found by adding the mass between finite elements and interpolating at the boundaries where the concentration spans the flammable limits. With such models, the level of detail and the "fineness" of the steps in the predicted concentration curves depends very much on the number and size of grid elements chosen.

Flash Fire Ignition Tests

Experiments were reported by Evans and Puttock (1986), that combined a realistically dispersing cloud of propane and an ignition source that varied in distance to the cloud. These tests can be used to verify flammable mass predictions by models.

Eight experiments were conducted at Spadeadam, UK by releasing propane at a constant rate into a water-filled diked area in the corner of a level concrete pad 100 m on edge. A moveable cart carried five sets of igniters and gas concentration sensors, all at an elevation of 0.5 m. The cart was moved toward the propane source until the propane ignited as detected by an infrared camera. Ignitions were classified as "small flames" if a momentary flame occurred with a zone of hot gases 4 to 16 m diameter. "Large flames" were momentary with a hot zone greater than 16 m. "Sustained flames" would travel back to the source. The experiment was, in effect, a large-scale flashpoint determination.

Successful tests were those in which an ignition occurred that met the above definitions. For the successful tests, the spill rate was 4.9 to 5.85 kg/s and wind speeds ranged from 2 to 7 m/s. "Sustained flames" were reported for downwind distances from 55 m to 79 m. The release Richardson number ranged between 3 and 110, indicating that heavy cloud effects were important. Richardson numbers were calculated using assumed ambient conditions (15 C, 1 atm., 75% humidity, D stability, 2 cm roughness length) and a propane boil-off mass flux of

0.12 kg/(m^2s). In similar experiments using the same type of gas sensors Colenbrander and Puttock (1984) reported that the mean concentration below which "sustained flames" are not obtained is 0.9LFL. (For propane the LFL is 2.1% and 0.9LFL is 1.89 %.)

Figures 5.6 and 5.7 plot centerline concentration profiles in the downwind direction for Spadeadam tests at wind speeds of 7 m/s and 3 m/s. Observations are given as concentration and distance points for four levels of ignition/nonignition. The observations are compared with predictions by the PHAST and HEGADAS models. The mean concentration profile predicted by HEGADAS is low in both cases. However, Evans and Puttock suggest that mean concentrations should be multiplied by an arbitrary factor of 1.4 to obtain peak concentrations which would ignite. The dashed lines in Figures 5.6 and 5.7 plot 1.4 times the mean concentrations predicted by the HEGADAS model and is a better fit of the observations than the mean concentration curve. At the high wind speed, the mean concentrations predicted by PHAST are acceptable in Figure 5.6 and slightly high in Figure 5.7.

Predicted flammable mass is compared with the Spadeadam experiments by Spicer and Havens (1996) for the DEGADIS and FEM3 models. They report that for a discharge rate of 5.4 kg/s the DEGADIS transient model predicts that the flammable cloud will reach steady-state conditions in 30 s for a 7 m/s wind speed, and in 120 s for a 2 m/s wind speed. This is illustrated for a 2 m/s wind speed in Figure 5.8. Figure 5.8 shows DEGADIS model predictions for the mass

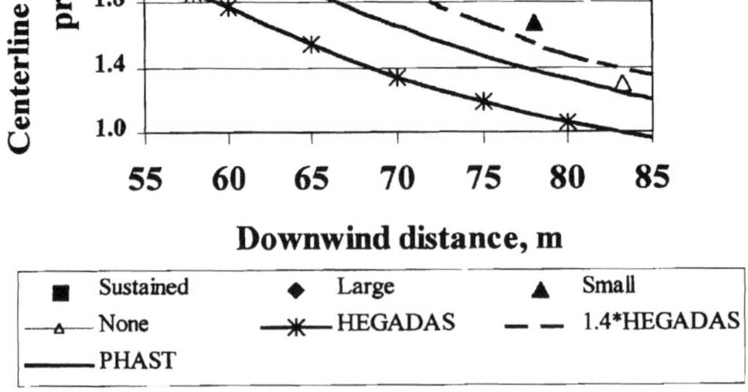

Figure 5.6. Comparison of PHAST and HEGADAS predictions with Spadeadam Test 202 at 7 m/s wind (data from Evans and Puttock, 1986).

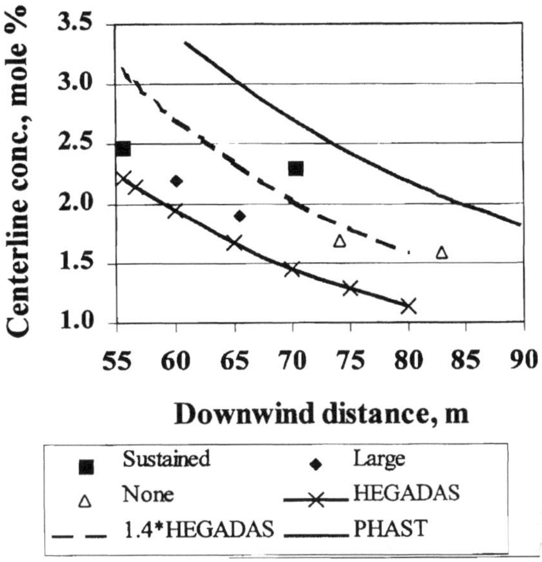

Figure 5.7. Comparison of PHAST and HEGADAS profiles with Spadeadam Test 203 with 3 m/s wind (data from Evans and Puttock, 1986).

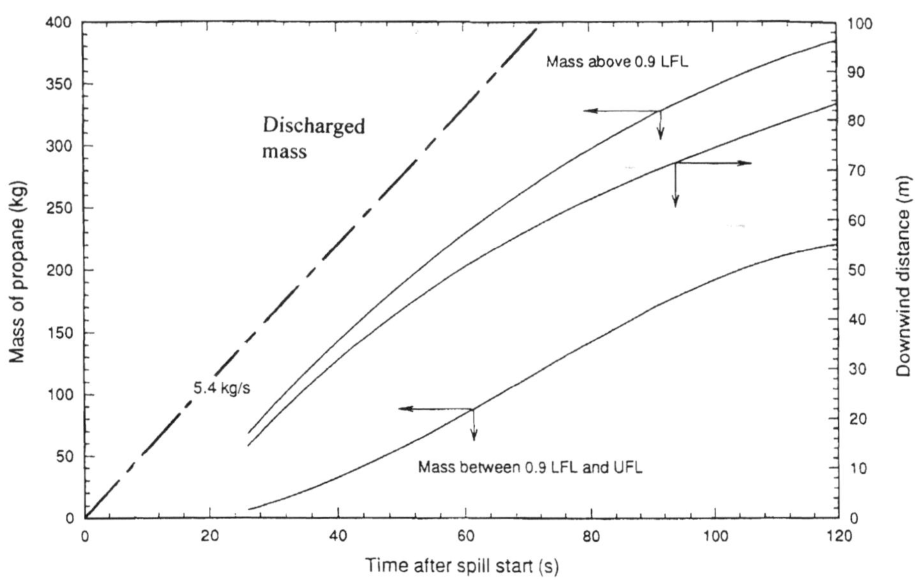

Figure 5.8. Build up of flammable mass of propane from Spadeadam tests at low wind speed (2 m/s) as predicted by the DEGADIS model (Spicer and Havens, 1996). (Reproduced by permission of Elsevier Science Publishers.)

above 0.9 LFL and between 0.9 LFL and the UFL. As would be expected, the mass above 0.9 LFL increases at early times nearly parallel to the discharge rate curve, and in later times it begins to level out to reach a steady-state. The leveling out is not complete, though, showing that the choice of 120 s. as the time to reach steady-state is somewhat arbitrary. The difference between the upper curve (mass > 0.9 LFL) and the lower curve (mass between 0.9 LFL and UFL) is, of course, the mass above the UFL, showing that in this example about half of the propane in the cloud is above the UFL. The flammable mass (0.9 LFL to UFL) goes from 14% at 40 s to 34% at 120 s.

Predictions of the DEGADIS, FEM3, and PHAST models for the Spadeadam test conditions are summarized in Table 5.1A. The agreement for these models is good except for FEM3 predictions at the lower wind speed. Table 5.1B shows that flammable mass predicted by both DEGADIS and FEM3 decreases strongly with increasing wind speed. In Table 5.1B the FEM3 predictions are considerably higher than those of DEGADIS, especially at low wind speed. Since the FEM3 predictions for flammable extent are higher than observations at low wind speed, the DEGADIS predictions are more believable. Recognize, these conclusions are obtained from only one field experiment.

Analytic formulas to predict flammable mass were developed by Hesse (1991) as shown in Appendices I and K. Hesse's formulas integrate Gaussian profiles, and so apply only for passive dispersion. Spicer and Havens (1996) also

TABLE 5.1A
Comparison of Three Models' Predictions of Distance to Ignitable Concentration with Spadeadam Tests

Wind Speed at 10 m (m/s)	Observed Distance to Sustained Ignition (m)	Extent to 0.9 LFL (m)		
		By DEGADIS	By FEM3	By PHAST
2	79	80	180	87
7	55	47	65	64

TABLE 5.1B
Comparison of Two Models' Predictions of Flammable Mass with Spadeadam Tests

Wind speed at 10 m (m/s)	Maximum Mass of Propane above 0.9 LFL (kg)		Maximum Mass of Propane between UFL and 0.9 LFL (kg)	
	By DEGADIS	By FEM 3	By DEGADIS	By FEM 3
2	390	1410	220	990
7	47	86	35	56

compared the formulas of Hesse with those of the DEGADIS model. The conditions of the Spadeadam tests were used except that to make the comparison more appropriate to the assumption of passive dispersion the discharge rate of propane was reduced until the release Richardson number decreased to 1.0 (at 0.8 kg/s discharge rate).

Table 5.2 gives the results at 0.8 kg/s and at 0.4 kg/s discharge rate. Even so, near the source the vapor cloud will be heavier than air and the release Richardson number will be above 1.0 (so passive models are not appropriate). The predicted extent to the 0.9 LFL concentration is slightly longer by DEGADIS. The maximum flammable mass between the UFL and 0.9 LFL by DEGADIS is about twice as high as Hesse's formulas predict. Yet, Hesse's formulas predict much more flammable mass above 0.9 LFL than does DEGADIS. The DEGADIS predictions imply a small fraction of the mass is above the UFL, whereas Hesses's predictions say that a much larger fraction is above the UFL. This implies that the entrainment rate is lower for Hesse's model which uses passive entrainment than for the DEGADIS model for which entrainment is by both dense gas and passive mechanisms. The inconsistency here is that heavy gas entrainment rates are theoretically lower than passive entrainment rates. The applicability of passive models, already in doubt, has further cause for suspicion.

The flammable mass is the mass in each puff after it dilutes to the UFL and before it dilutes to the LFL (typical of top hat models). Modeling dispersion with only the passive dispersion or heavy gas dispersion mechanisms would greatly underpredict the air entrainment rates (dispersion) and overpredict the flammable mass in such a cloud.

Treating Highly Transient Discharge with Successive Puff Model

Some discharge rates are highly transient, such as occur when a pressure relief valve opens or a rupture disk breaks on a high-pressure system. For example, Figure 5.9 shows how pressure and discharge rate fall off very rapidly when a rupture disk opens on a high-pressure polyethylene reactor (Rota et al., 1995). With such a highly time-dependent, momentum-dominated puff, the flammable mass is best estimated using the method described next.

TABLE 5.2

Comparison of Hesse's Method and DEGADIS for Passive Dispersion Propane Releases

Release Rate (kg/s)	Maximum Distance to 0.9 LFL (m)		Maximum Mass of Propane above 0.9 LFL (kg)		Maximum Mass of Propane between UFL and 0.9 LFL (kg)	
	Hesse	DEGADIS	Hesse	DEGADIS	Hesse	DEGADIS
0.4	11	16	5.7	1.1	0.4	0.9
0.8	17	21	17	3.0	1.2	2.4

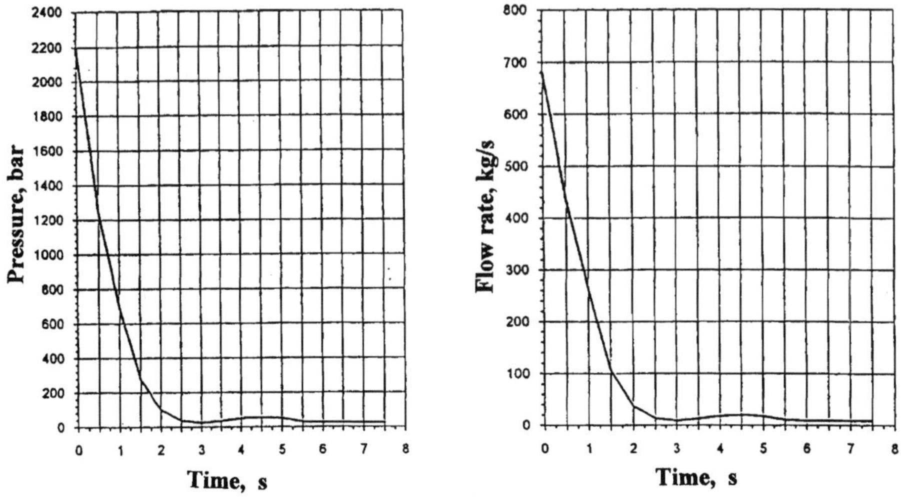

Figure 5.9. Highly transient pressure and discharge rate from high pressure polyethylene reactor after rupture disks open. (Reproduced by permission of Elsevier Science Publishers.)

A commonly used method for treating highly transient releases is to divide the discharge rate into small time increments and model each increment as an instantaneous puff (*Yellow Book*, 1997, p. 4.46). For example, Rota et al. (1995) used a top hat puff model to produce the cloud profiles between the LFL and the UFL illustrated in Figure 5.10 . This example breaks the discharge rate of

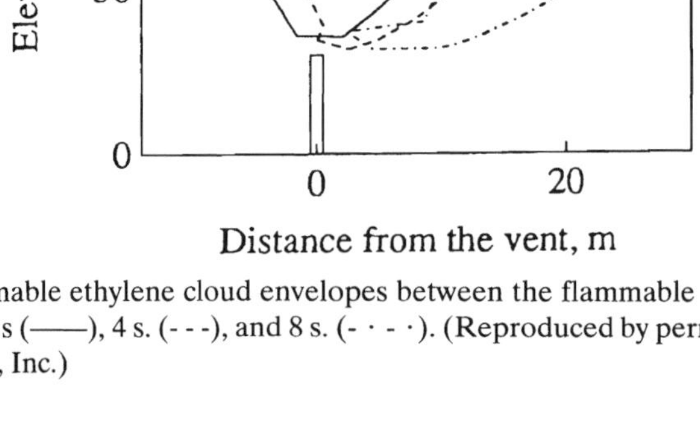

Figure 5.10. Flammable ethylene cloud envelopes between the flammable limits (Rota et al., 1995). Profiles at 2 s (——), 4 s. (- - -), and 8 s. (- · · ·). (Reproduced by permission of Elsevier Science Publishers, Inc.)

ethylene in Figure 5.10 into a number of puffs of duration <1 s each, models each puff, and assembles the envelope of puff widths at the same moment in time, in this case 2 s, 4 s, and 8 s. At 2 s after the release, the wind has little influence on this high-pressure discharge. At 4 s, the size of the flammable cloud has decreased considerably since the leading edge of the puff dilutes rapidly. After 8 s, the flammable cloud envelope is started to be advected by the mild wind speed of 2 m/s. In this case, the jet momentum mechanism is dominant, so cloud buoyancy is not relevant.

Flammable Mass from Passive Dispersion Models

Passive dispersion is dominated by normal atmospheric turbulence and applies where the plume is dilute or neutrally buoyant (plume density equals ambient air density). For flammable materials, most compounds have a molecular weight greater than that of air, so only if the released material is warm could the release be neutrally buoyant. Furthermore, a dilute cloud would generally be considered one with a concentration less than about 3%. Some materials have an LFL above 3%, so these would not be dilute enough that dispersion would be passive by the time they reach the LFL. Even so, the passive model is often invoked, as an approximation.

The passive dispersion model is an analytic solution for the partial differential equations describing point-source or line source diffusion, generalized to turbulent diffusion. The solution in this regime gives concentration profiles as normal distributions summarized in Appendix I. The centerline concentration decreases as the cloud width increases in going downwind as depicted schematically in Figure 5.11 (Schumann and Wirkner-Bott, 1991).

The dispersion or air entrainment rate for a Gaussian model is expressed inherently by the increasing cloud dimensions downwind. Cloud width is expressed by the lateral standard deviation $\sigma_y(x)$ and cloud height by the vertical standard deviation $\sigma_z(x)$. The standard deviations vary only with downwind distance, atmospheric stability class, and surface roughness length. Different formulas are in use relating $\sigma_y(x)$ and $\sigma_z(x)$. Two alternative forms are used in Appendices I, J, and K. Appendix I lists the basic Gaussian solutions applicable to instantaneous and continuous releases, point and line sources. It also provides a rigorous solution for flammable mass for an instantaneous, point source. Appendix J provides an approximate solution for a continuous source. Appendix K gives a rigorous solution for a continuous source. The methods developed in these appendices are summarized below.

Flammable Mass for Passive Dispersion Model—Instantaneous, Point Source

Appendix I treats an instantaneous release as a point source. The solution in Appendix I given as Eq. (I.12) gives the maximum ratio of flammable mass to

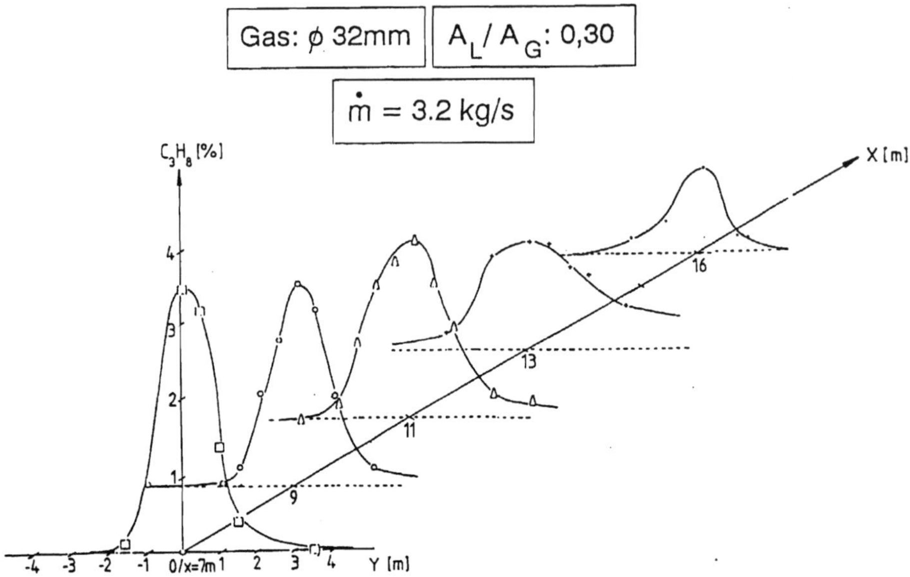

Figure 5.11. Visualization of successive concentration profiles in the Gaussian modeling regime.

total mass of fuel in a cloud. The solution for an initial concentration above the UFL gives the ratio of flammable mass to total mass of fuel in the cloud:

$$\frac{m_{\mathrm{f}}}{m}=\operatorname{erf}\left[\sqrt{\ln\left(\frac{c_0}{c_{\mathrm{LFL}}}\right)}\right]-\operatorname{erf}\left[\sqrt{\ln\left(\frac{c_0}{c_{\mathrm{UFL}}}\right)}\right]$$
$$-\frac{2c_{\mathrm{LFL}}}{c_0\sqrt{\pi}}\sqrt{\ln\left(\frac{c_0}{c_{\mathrm{LFL}}}\right)}+\frac{2c_{\mathrm{UFL}}}{c_0\sqrt{\pi}}\sqrt{\ln\left(\frac{c_0}{c_{\mathrm{UFL}}}\right)} \tag{5.7}$$

A simplification of Eq. (5.7) is plotted in Figure 5.12. For typical materials such as propane, the ratio of UFL/LFL is 5, so by Figure 5.12 approximately 60% of the mass of fuel in the cloud is in the flammable region. Equations (I.9) to (I.12) apply with any formulation for σ_x, σ_y, and σ_z as a function of x.

Flammable Mass for Passive Dispersion—Approximate Method

The *Yellow Book* (TNO, 1997, pp. 4.68–4.76) describes an approximate solution for the distance to the LFL and UFL. These approximate solutions can be used in an expression for flammable mass of continuous releases (both point source and line source).

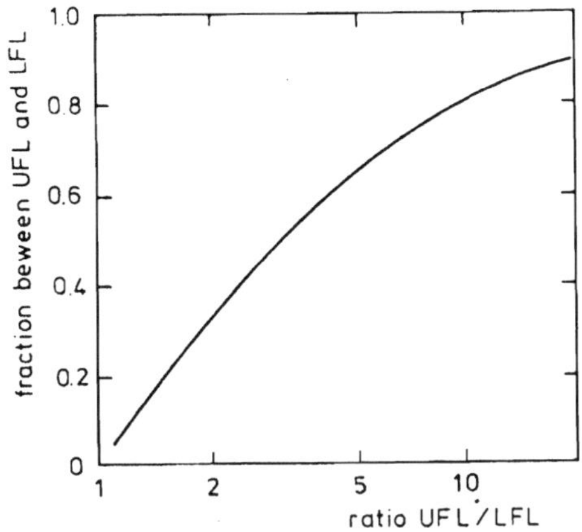

Figure 5.12. Maximum mass fraction of gas in the flammable region for an instantaneous release (Van Buijtenen, 1980). (Reproduced by permission of Elsevier Science Publishers, Inc.)

The approximate formula for flammable mass between the UFL and LFL for a continuous point source is given in terms of the wind speed, u_w, and the distances to the flammable limits, x_{LFL}, x_{UFL}, by the dimensionless expression:

$$\frac{f_m}{w} = \left[\frac{f_{z2}(L)+1}{f_{z2}(L)+2}\right]\left[\frac{x_{LFL} - x_{UFL}}{u_w}\right] \tag{5.8}$$

Here, L is the Monin–Obukov length, discussed in Appendices B and J. The function f_{z2} is given by Eq. (J.6) in Appendix J, and restated below

$$f_{z2}(L) = 0.249 \arctan\left[C_5(1/L + C_6)\right] + 1.041 \tag{5.9}$$

The values of the dimensional constants are $C_5 = -214$ m; $C_6 = 0.00077$ m^{-1}.

Example Plots of Flammable Mass for Gas and Liquid Releases

Example calculations are provided in Figures 5.13 through 5.18 for typical gas and liquid hydrocarbon releases. These plots show the sensitivity of flammable mass predictions to the factors affecting discharge rate, hole size and system pressure.

The direction of the release becomes very important to the cloud formation as illustrated in Section 7.1. The dispersion of a high pressure nonbuoyant gas jet is basically driven by the velocity of the gas, because it is so much larger than typical wind speeds. Neither does atmospheric stability have much of an effect on the dispersion of a high-pressure gas jet release. In fact, for jet releases the flammable mass is proportional to the length of the plume, L_i. The plume length, in turn, is proportional to the discharge rate, which is proportional to the molecular weight, the area of discharge and the square root of the pressure difference:

$$L_j \sim M_t P^{1/2} * D_0^{\,2}$$

Figure 5.13 gives the release rate for an ideal gas chart of molecular weight 28, $\gamma = 1.4$ (Carbon monoxide). The PHAST program using the Peng–Robinson equation of state modeled the dispersion giving the results listed in Table 5.3 for a carbon monoxide discharge. This illustrates the insensitivity of the flammable mass to wind speed and atmospheric stability class. The model was further applied to produce Figure 5.14 for carbon monoxide releases. Figure 5.14 shows that equivalent hole diameters of 2 in. or larger, the flammable mass in a jet falls along a single curve. For smaller hole sizes, the flammable mass curves drop

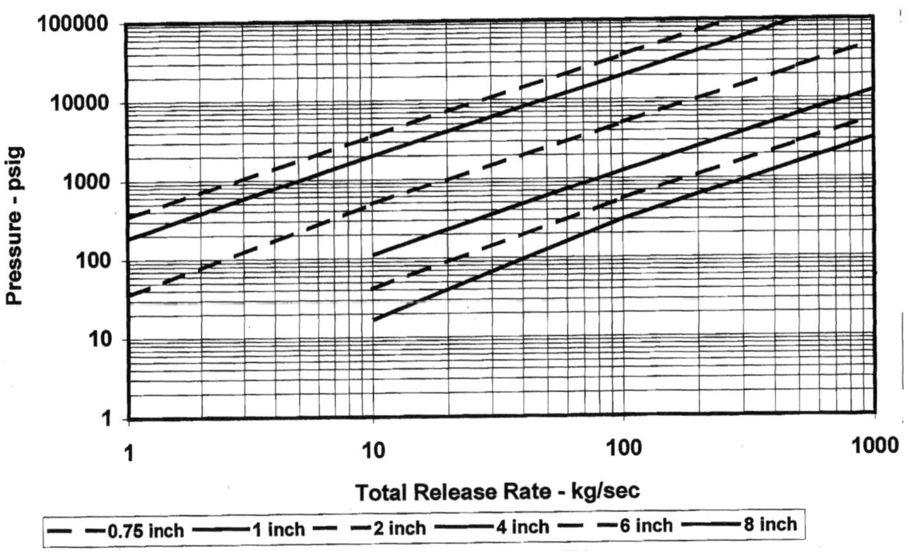

Figure 5.13. Ideal gas discharge rate as a function of hole size and vessel pressure for a release of gaseous carbon monoxide.

TABLE 5.3
Example Calculation of Flammable Mass for a Gas Jet of Carbon Monoxide

Extent of Gas Jet (feet) to ½ LFL for Various Wind Speeds, Atmospheric Stabilities		
3.4 mph; F	11.2 mph; D	0.2 mph; D
72	72	76

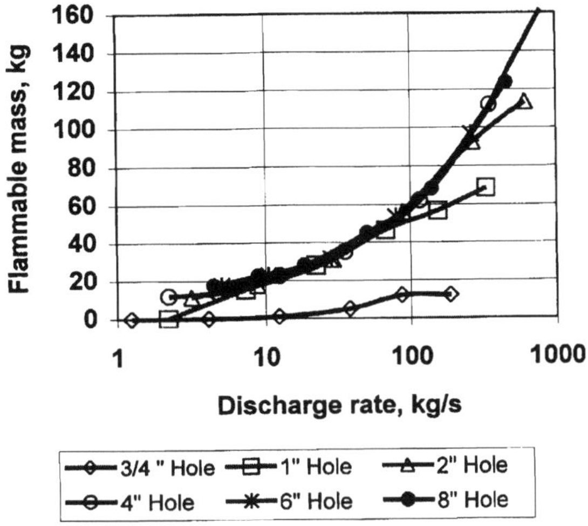

Figure 5.14. Flammable mass as a function of discharge rate for a release of gaseous carbon monoxide.

below the curve for larger diameters. The single curve for larger hole sizes is a consequence of the self-similar nature of jets. The discharge rate is proportional to the discharge area (that is the square of the hole diameter), and the air entrainment rate is proportional to the momentum of the jet which is the mass in the jet times its velocity. The mass in the jet is proportional to the discharge rate (and discharge area).

Example 8. Flammable Mass for Gas Release

Find the flammable mass from a horizontal release of CO from a 500 psi container having a 2-in. diameter hole. From Figure 5.13 the discharge rate is 10 kg/s. From Figure 5.14 the flammable mass is about 20 kg. Notably, nearly the same flammable mass occurs at a given discharge rate. That is, the same discharge rate and flammable mass occur with a small hole size and a large pressure as with a larger hole size and a smaller pressure.

Liquid releases are treated in Figures 5.15–5.18. Discharge rate is plotted in Figure 5.15 for various hole diameters and operating pressures for liquid releases that are either subcooled or flash after release. The graph uses the equations in Section 4.1.1. Figure 5.16 plots the flammable mass as a function of release rate, according to the TNO approximation method. Figures 5.15 and 5.16 are appropriate for liquid hydrocarbon releases (e.g. propane, butane) with at most low superheat (<10 C). For most cases the predictions shown are conservative, due to the following assumptions.

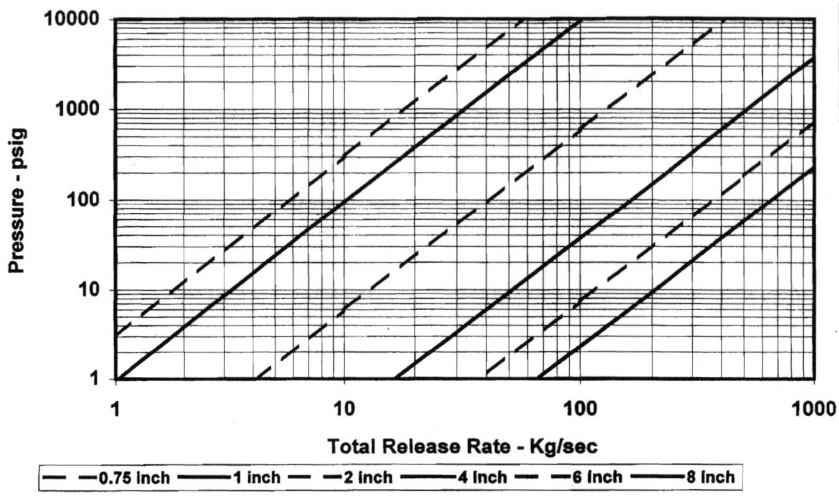

Figure 5.15. Discharge rate as a function of hole size and vessel pressure for liquid hydrocarbon.

Assumptions

Material: Light flashing hydrocarbon above its atmospheric boiling point

Discharge coefficient, C_D	= 0.61
Liquid density, ρ_L	= 800 kg/m^3
UFL	= 10 mole %
LFL	= 2.0 mole %
Release height	= 1 m
Wind speed	= 1.5 m/s
Stability class	= F
Roughness length	= 1 m
Wind profile reference height	= 10 m
Inversion height	= 10000 m

The cloud size to 0.9 LFL with the above assumptions is graphed in Figure 5.17. The cloud for a continuous, elevated release, according to the Gaussian model, is an elliptical shape in the vertical plane. After touchdown, it will be a semi-ellipse in the vertical plane, but this case is not treated in the example plots. The cloud contours according to Figure 5.17 can be overlaid on a plant plot plan to determine what part of the cloud would be in congested areas. For simplicity, the entire cloud is assumed to occur in 2D confinement with a medium obstacle density, for a fuel of medium reactivity. Explosion overpressure contours predicted by the Baker–Strehlow method are graphed for these assumptions in Figure 5.18. The average heat of combustion of 47 MJ/kg is used for this calculation.

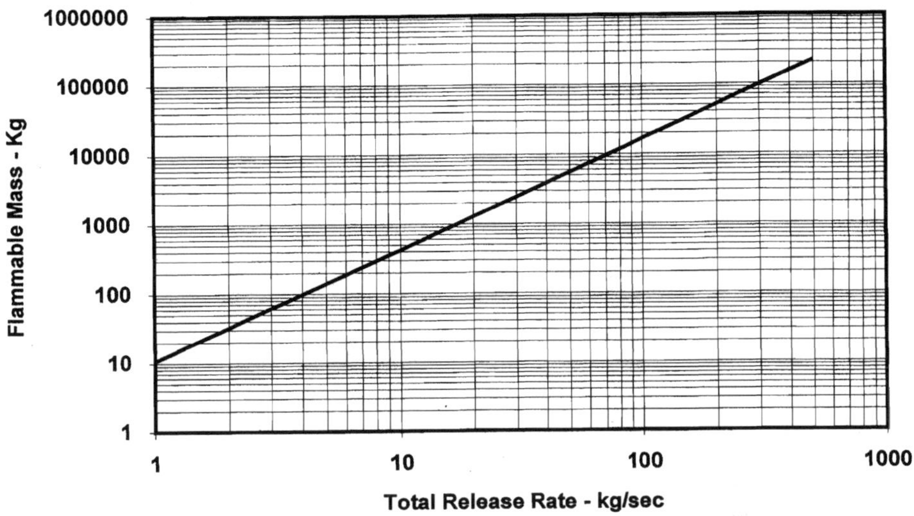

Figure 5.16. Flammable mass between 0.9 LFL and UFL (F stability; 1.5 m/s wind speed) predicted by TNO approximate method as a function of discharge rate for a liquid hydrocarbon release.

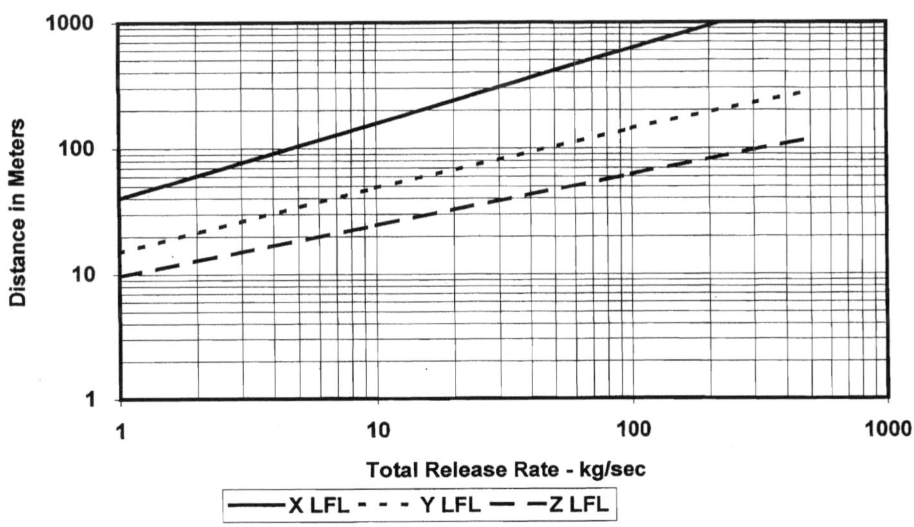

Figure 5.17. Vapor cloud dimensions to flammable limit concentrations as predicted by the TNO approximate method (F stability; 1.5 m/s wind speed).

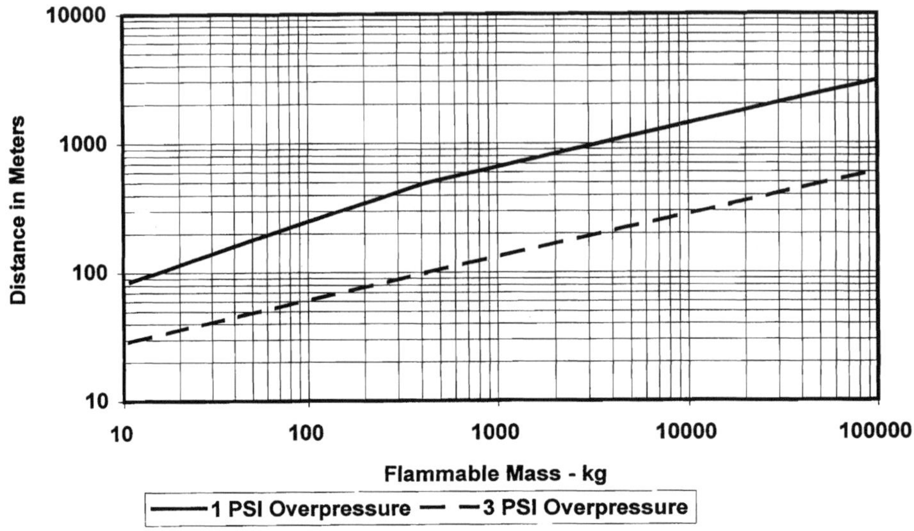

Figure 5.18. Distance to specified peak overpressures for explosions in typical plant conditions as a function of flammable mass, as predicted by the Baker–Strehlow method.

Example 9. Flammable Mass for Liquid Release

Estimate the flammable mass for an accidental release (e.g., from an overhead drum from distillation unit) of an average light liquid hydrocarbon within a process unit having the example case configuration (2D, medium congestion). The equivalent hole diameter is estimated to be 2 in. (0.0508 m). The operating pressure is 100 psig (6.9 bar(g)). From Figure 5.15, the discharge rate is 40 kg/sec. This leads to 4000 kg flammable mass, as read from Figure 5.16.

The cloud size is 350 m in the x direction (long), $90 \times 2 = 180$ m in the y direction (wide) and 40 m in the z direction (high), as read from Figure 5.17. The 4000-kg flammable mass assumes that this entire cloud is confined within the process unit, or units.

Figure 5.18 gives a distance of 200 m to the 3 psi overpressure and approximately 1000 m to the 1 psi overpressure, from the center of the explosion which is taken to be the center of the cloud (175 m from the release).

Flammable Mass for Passive Dispersion Model—Rigorous Solution

A rigorous solution for the flammable mass of a continuous release Gaussian model is given by Hesse (1991) in terms of standard tabulated mathematical functions, elliptic integrals of the second kind, $E(k,\theta)$ and an infinite series known as confluent hypergeometric functions. (These are solutions of a special class of differential equations like the more familiar Bessel functions.) His solution applies for the simpler power law form of the σ_y, σ_z curves:

$$\begin{aligned}
\sigma_y &= a(x + x_v)^m \\
\sigma_z &= b(x + x_v)^n \\
\sigma_{xp} &= c(x + x_v)^j \\
\sigma_{yp} &= d(x + x_v)^k \\
\sigma_{zp} &= c(x + x_v)^l
\end{aligned} \tag{5.10}$$

where x_v is the adjustable virtual source distance.

The value of having semianalytic solutions, even though they require considerable effort to program, is that the approximation error in these solutions is readily estimated and thus can be well controlled. These solutions, then serve as a check on the numerical integration techniques described previously.

5.3. Methods for Finding the Flammable Mass in Partially Confined Vapor Clouds

Vapor clouds discharged in process plants often pass through areas with various degrees of congestion. A commonly accepted way of treating such releases is to use an "unconfined" dispersion model to produce estimates of the vapor cloud footprint and overlay this footprint on the process layout diagram as illustrated in Figure 5.19 (Baker et al., 1997). To take account of the enhanced turbulence

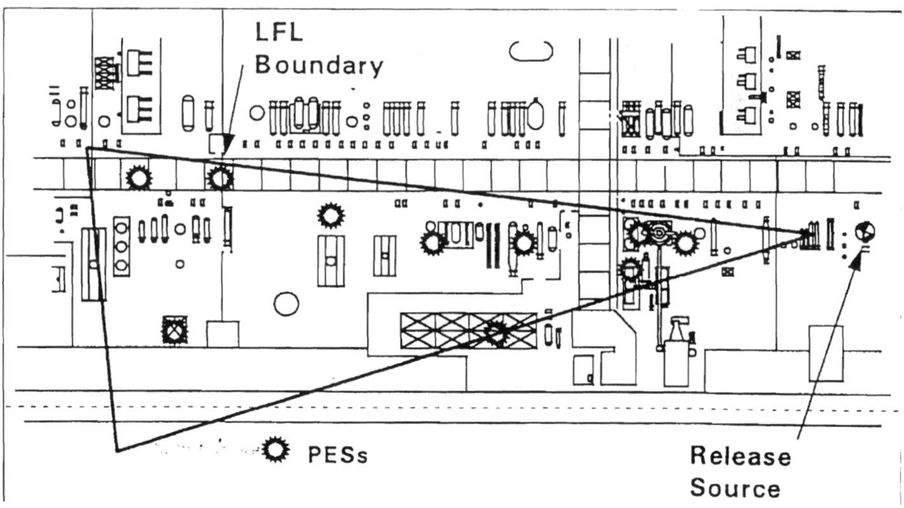

Figure 5.19. Example lower-flammable-limit footprint of vapor cloud in process area (from Baker et al., 1997).

effect of process equipment, the surface roughness length is increased in a process area as is discussed in Section 4.2.3. The example plot shows several areas within the flammable vapor cloud that are potential explosion sites (PES), or areas with a high degree of confinement and/or congestion and which have a reasonable probability of ignition. The following subsections detail how to estimate the mass of flammable material in one or more PES areas. This calculation is put in context by also detailing the steps of methods to estimate explosion effects.

5.3.1. Estimating Flammable Mass for Potential Explosion Sites

Section 4.8.4 lists the steps in applying the volume source multi-energy and Baker–Strehlow methods. These steps are illustrated in Chapter 7. The procedure pertinent to estimating flammable mass is summarized here.

The first step in the volume-source methods is to map on the plant plot plan rectangular areas with a fairly uniform degree of confinement and congestion. As detailed later, congestion is evaluated as low, medium, or high. Areas with medium or high congestion are considered PESs even if the explosion is allowed to expand in three dimensions. According to the methods, adjoining PESs should be consolidated to make a single PES using area-averaging of their confinement and congestion parameters. PES areas which are close to each other should be consolidated as well, as discussed in Section 4.8.8.

In steps 2, 3, and 4 of the volume-source methods a vapor cloud dispersion model is used to predict the concentration profiles, density gradient, and effective cloud height. The flammable cloud footprint may completely or partially overlap a PES. In either case, find an average cloud height and density over the cloud area within each PES. That is, designating A_{mi} and h_{mi} as the area and height of the flammable cloud within the ith PES. The formula requires an average concentration of flammable material over the PES, c_{mi}, in kg/m^3 (lb/ft^3). This can be estimated from the profiles developed from the dispersion model. The corresponding flammable mass, f_{mi}, is found with

$$A_{mi} = \text{minimum}(A_{PESi}, A_{cldi}) \tag{5.11a}$$

$$h_{mi} = \text{minimum}(\bar{h}_{PESi}, \bar{h}_{cldi}) \tag{5.11b}$$

$$c_{mi} = \frac{\iint c(x,y)\,dy\,dz}{A_{mi}} \tag{5.11c}$$

$$f_{mi} = A_{mi} h_{mi} c_{mi} \tag{5.11d}$$

where A_{PESi} and $\bar{h}_{PESi}$ are the area and confined height of a 1D or 2D PES (for a 3D PES the height is infinite), and A_{cldi}, $\bar{h}_{cldi}$ are the area and average height of a flammable cloud within the PES.

Step 5 of the volume-source methods is to estimate the average heat of combustion over each PES accounting for differences in cloud composition for multicomponent clouds. That is, obtain a weighted average of the heat of combustion for the cloud in each PES, H_{ci}, in J/kg and

$$H_c = \sum_i f_{mi} H_{ci}(c_{mi}) \tag{5.12}$$

In practice, the volume-source explosion prediction methods do not require precise averages as were found for unconfined clouds in Section 5.2. In fact, the original form of the multi-energy method suggests using as the average concentration the stoichiometric concentration. The original method further suggests it is tolerable to make use of the approximation that the heat of combustion per unit volume at stoichiometric air–fuel concentration is nearly constant for most hydrocarbons, at H_{cv} = 3.5 MJ/m^3 (Mercx and van den Berg, 1997) giving

$$H_{comb} \cong H_{cv} A_{mi} H_{mi} \tag{5.13}$$

To calculate H_{cv} from the values in Appendix C use the mass fraction of combustibles, x_c and the cloud density, ρ_v:

$$H_{cv} = H_c x_c \rho_v \tag{5.14}$$

Steps 6 through 10 of the volume-source explosion methods are illustrated in Chapter 7 for specific examples.

5.4. Methods for Finding the Flammable Mass in Confined Vapor Clouds

Only a small amount of flammable mass released inside a building can produce a damaging explosion. As an example, the potable water supply of a chemical plant was contaminated by a hydrocarbon stream composed of pentanes and aromatics (benzene, toluene, xylenes). In the process of purging the hydrocarbons from the water supply the showers were left running in the operators' locker building. As estimated 9 kg (20 lb) of hydrocarbon built up in the building by evaporating from the shower water. This was enough to reach the LFL. Some 10 hours or so later, when the air conditioner turned on, a spark ignited these vapors and a weak explosion blew out the transite walls at two opposite ends of the building (Marshall, 1998).

Calculating the flammable mass for an indoor release must consider possible barriers and stratification. Stratification develops when a spill of heavy gas or evaporating liquid develops a substantial vertical concentration gradient, or light gases form a gradient from the ceiling downward. A scenario of concern for indoor explosions is a small spill of volatile solvent such as acetone or paint thinner. The evaporation such a spill can produce a flammable vapor cloud that remains stratified by negative buoyancy or localized by obstacles that act as effective partitions. A leaking natural gas line in a room with a circulating ventilation system will usually produce a well-mixed vapor cloud filling the room. Alternately, we show later that when the room is quite large and the discharge velocity is low to moderate, even a gas discharge can build up a stratified gas layer. Stratification is important since part of the cloud can be in the flammable region, even when the same amount of material in a well-mixed room would be above the UFL or below the LFL.

A deflagration in a room filled with flammable vapor, aerosol, or dust is referred to as a full-volume deflagration (FVD). When the flammable cloud or dust occupies only part of the room volume the corresponding term is a partial-volume deflagration (PVD).

Even the strongest buildings can only withstand low internal pressures, typically on the order of 1–2 psi maximum overpressure. For an unvented deflagration, this pressure is generated by burning about 2–5% of the room volume (as a flammable gas). As a consequence, if explosion vents are to be effective, they must open very early in the incident. A major difference between an FVD and a PVD is that, depending on the relative location of the ignition source and the vents, most of the material discharged from an FVD could be unburned gas. Whereas for a PVD, with only a small portion of the building in the flammable region, all of the flammable gas could burn inside.

In practice, the unburned material discharged through explosion vents will be extremely turbulent and will see a strong ignition source (the flame behind). Thus, an external explosion can result which could produce higher overpressures than the initial pressure in the vented building. For example, Monsanto Chemical Co. conducted a test in which the overpressure from the external explosion was about an order of magnitude greater than predicted (Birtwisle, 1998).

Since explosions from an FVD are almost invariably damaging to an intolerable degree, attention has focused on mitigation of indoor explosions by designing explosion venting panels. Good practice in keeping with the "layers of protection" concept is to install flammable gas detection systems to automatically initiate corrective action, backed up by vent panels.

In an unvented FVD the maximum pressure occurs immediately before the flame front reaches the walls. The deflagration pressure is felt almost simultaneously on all the internal surfaces. The theoretical overpressure of an unvented FVD equals the expansion ratio for a material discussed in Section 5.3.4. This ranges from six to nine and is typically around eight times the initial absolute pressure. The enclosure strength sets a tolerable reduced pressure level, ΔP_{red}, which is the design objective of explosion vent design, and is typically 0.2 bar(g) or less. Clearly, to keep internal pressures below the ΔP_{red} set-point, the vents must open rapidly. Hence, the inertia of vents can significantly affect the dynamics of venting and the maximum overpressure inside the building.

The design of explosion venting panels for gases and dusts is treated in NFPA 68 (1994) and in a series of papers by Tamanini (1995, 1996a, 1996b, 1996c, 1998). The vent design methods favored in Europe are outlined in VDI 3673 (1995). The NFPA 68 methods parallel the VDI 3673 methods.

Tamanini (1996b) extended designed correlations from FVDs to PVDs by making use of the fill fraction of a PVD, X_r, defined as the volume of the vapor cloud divided by the volume of the room.

Design correlations for the area of explosion vent panels, A_v, give this area as a function of the allowable maximum reduced overpressure, ΔP_{red}, which occurs shortly after the vent panels open. An example correlation is that of Tamanini (1995) for dust explosions, but the principles are expected to apply as well to gas explosions. This is shown in Figure 5.20 as the dashed line. The ordinate is the normalized pressure ratio or $\Delta P_r / \Delta P_m$. For PVDs the ordinate is divided by the filled fraction of the room, X_r. The abscissa values are the correlating parameter defined in terms of a constant, a_{cd}, the room volume, V, the theoretical maximum overpressure, the vent area, and a reactivity parameter, K. For PVDs the abscissa is multiplied by the fill fraction, X_r

$$\Gamma = a_{cd} \frac{A_v}{V^{2/3}} \frac{\Delta P_m}{K_{st}} \tag{5.15}$$

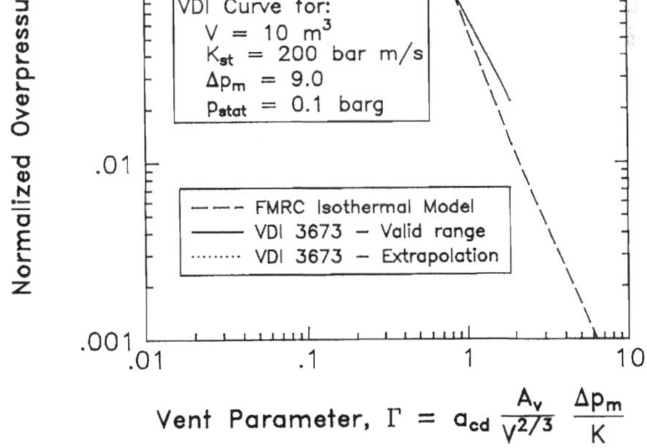

Figure 5.20. Example correlation to design explosion vent area as a function of the maximum reduced overpressure normalized to the theoretical maximum overpressure. (Reproduced by permission of Factory Mutual Research, Inc.)

The dust reactivity, K_{st}, and a corresponding gas reactivity, K_g, are found by explosion experiments in which pressure is recorded and the maximum pressure rise slope is found. The constants K_{st} and K_g are defined as

$$K = V^{1/3}\left(\frac{dP}{dt}\right)_{max} \tag{5.16}$$

The straight line segment slightly above the dashed curve is the VDI 3673 correlation for the same conditions as the dashed line. This example is chosen to show that the correlations given by NFPA 68 and VDI 3673 are straight-line segments corresponding to a more general curve and this general curve reaches the maximum indoor overpressure when the vent area is zero.

Given a vent area, the vent panel correlations can be solved for the corresponding reduced overpressure, so they act as an explosion model, albeit at only a single distance. The more recent vent sizing correlation extensions (Tamanini, 1998) include correlations for the blast wave duration (hence also impulse).

NFPA 68 and VDI 3673 provide a method to design explosion vents for "low strength enclosures" such as normal buildings. It is based on indoor tests from various sources, and analysis of actual building explosions. In practice, the pressure it predicts is signficantly less than would be calculated using the Baker–Strehlow or multi-energy methods for a highly congested structure (with no walls). The discrepancy between the methods may reflect the fact that the turbulence levels in the tests used for indoor explosions may have been much higher than the turbulence levels inherent in outdoor tests that are the basis for the Baker–Strehlow and multi-energy methods. This topic is a subject for future developments as discussed in Chapter 8.

For some situations it is possible to calculate the concentrations, and hence the flammable mass inside a room or building. For one case, if the spill is outdoors, we can find the concentration indoors from a passing vapor cloud. For another case, if the spill occurs indoors and the room ventilation is turned off we can calculate the concentration profiles that build up indoors. A third case is the spill of a volatile liquid indoors. These cases are treated in the following subsections.

5.4.1. Flammable Mass in Well-Mixed Room from Spill Outdoors

A spill outdoors generates a vapor cloud that may pass by and over a building. The outdoor concentration of flammable material is provided by a dispersion model as before in Sections 5.2 and 5.3. The outdoor concentration passing a building is, in general, time-varying in concentration. The objective of modeling indoor concentrations is to find the circumstances under which a flammable concentration can occur indoors.

A generalized solution for indoor concentrations is given in Appendix G. This model allows for removal by a filter, removal by decomposition reactions, and removal by coating out on surfaces. For flammable materials, these removal mechanisms are usually not significant, so some terms may be dropped. Two special case solutions are shown in Figure 5.21 and given below.

For the simple cases in which the outdoor concentration is a "top-hat" form, illustrated in Figure 5.21, Eq. (G.1) reduces to

$$\frac{dc(t)}{dt} = \lambda[c_0 - c(t)] \tag{5.17}$$

where $\lambda = q_0/V$ is the reciprocal time constant $1/\tau$. Equation (5.17) has the solution

$$c(t) = \begin{cases} c_0[1-\exp(-\lambda t)] & \text{for } t < t' \\ \hat{c}\exp[-\lambda(t-t')] & \text{for } t > t' \end{cases} \tag{5.18a}$$

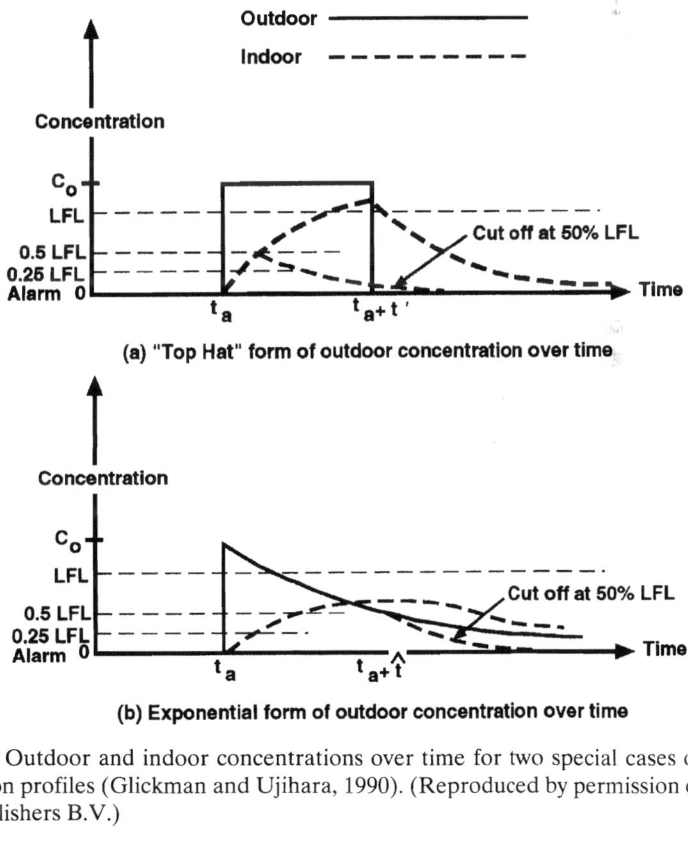

Figure 5.21. Outdoor and indoor concentrations over time for two special cases of outdoor concentration profiles (Glickman and Ujihara, 1990). (Reproduced by permission of Elsevier Science Publishers B.V.)

where

$$\hat{c} = c_0[1 - \exp(-\lambda t')] \tag{5.18b}$$

For the second special case the outdoor concentration is exponentially decreasing in time, or

$$c_0(t) = \hat{c}_0 \exp(-\mu t) \tag{5.19}$$

where $\hat{c}_0$ is the peak outdoor value that occurs when the cloud first arrives, and μ is the decay rate. This response occurs, for example, with a pipeline break (Wilson, 1979; Glickman and Ujihara, 1990). The solution is

$$c(t) = \hat{c}_0 \frac{\lambda}{\mu - \lambda}[\exp(-\lambda t) - \exp(-\mu t)] \tag{5.20}$$

which has its peak at $\hat{t}$ minutes after t_a, where

$$\hat{t} = \frac{\ln \mu_0 - \ln \lambda}{\mu - \lambda} \tag{5.21}$$

Shair and Heitner (1974) describe a similar model that relates indoor concentrations to outdoor concentrations The model allows for forced convection of fresh makeup air and recirculated air which generally pass through the same intake fan and filter. It also allows for spurious air infiltration through open windows, doors, and cracks in the building. Indoor concentrations are assumed to be well mixed by the action of the recirculation fan. Even when the fan is turned off, the assumption can be made of a well-mixed building since the intrusion rate of contaminant can only be high enough to become threatening (flammable) if the intrusion rate is substantial enough to cause mixing. No distinction is made between rooms in a building, except in special circumstances when the mixing between rooms is known to be limited.

Data on the air exchange rates of typical buildings are given by Grot and Persily (1986), Jann (1989), Purdy (1989), and Wilson (1991). In a survey, Wilson (1996) found that leaky houses have 2 air changes per hour. Normal office rates are typically around 3–4 air changes per hour. Restrooms and kitchens are typically around 6 changes/hour. Highly ventilated areas such as process areas and laboratories are typically designed for 12 air changes per hour. Emergency services are often designed to increase normal rates to18–24 air changes per hour. A maximum ventilation rate is considered 30 air changes per hour which is a threshold for tolerable noise and for encountering difficulty in opening doors.

The time-varying flammable mass indoors is simply the product of the room volume, V, and the time-varying concentration, $c(t)$, subject, of course, to $c_{LFL} < c(t) < c_{UFL}$.

$$m_f(t) = Vc(t) \tag{5.22}$$

Numerous mitigation measures are used to avoid indoor contamination. Among these are keeping a positive pressure inside, protecting the air intake by using a remote intake location and/or gas detectors, and using automatic ventilation shut off triggered by gas detectors.

5.4.2. Flammable Mass from Indoor Release in Well-Mixed Room with Low Ventilation

The previous section illustrates the importance of fresh air makeup rate on indoor concentrations from an outside source of flammables. The ratio of ventilation rate to room volume is the time constant for concentration build-up or decay. The case of indoor releases with high ventilation rates is complex and does not lend itself to an analytic solution. However, we are able to treat the simpler case in which the ventilation system is shut off while an indoor leak occurs. This case is interesting, because with a large-enough room or a small leak rate a stratified concentration profile can develop in the room. Again the flammable mass in the room is time-varying. We first show how to find the concentration profiles as a function of time. Integration across these profiles gives the flammable mass.

Experiments were conducted by Cleaver et al. (1994) and Marshall (1984) releasing natural gas and propane into a room of height, width, and length *H*, *W*, *L* with a small hole in the bottom edge. In the range of geometrical configurations considered, to a first approximation, the gas concentration in the enclosure was uniform across any given horizontal section, but varied in height. Typically, a well-mixed layer of constant depth, h_1, was produced with a stratified transition layer of depth h_2 growing beneath it (for lighter-than-air methane) or above it (for heavier-than-air propane). This is depicted in Figure 5.22, postulating a linear profile across the transition layer. The lower edge of the transition layer moves as the gas accumulates.

Marshall (1984) observed that buoyancy and jet momentum affects the depth of h_1. Buoyancy effects are evident since a release of methane aimed vertically upward produces a smaller well-mixed layer than the same release aimed horizontally. The largest well-mixed layer is produced if the release of methane is aimed downward. Changes in the horizontal position of the leak produced minor differences in the mixing behavior. However, significant differences arose when the height of the release, z_r, was varied. Momentum effects are evident since in the limit, sonic releases produce well-mixed clouds through the enclosure while subsonic releases tend to stratify. Marshall's observations were confirmed by Cleaver et al. using dye injection into a water tank. By extension, these observations apply in reverse for a release of heavier-than-air material.

A model based on the above observations is described in Appendix H. This requires numerical integration of the ordinary differential equations for the

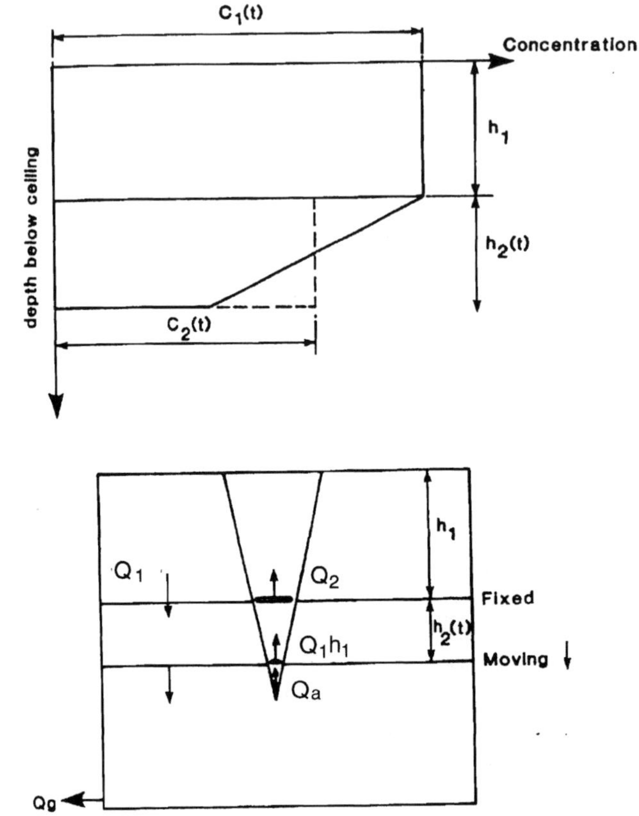

Figure 5.22. Idealization of stratification observed when buoyant methane gas is injected into a room from a small hole at the bottom of the room(Cleaver et al., 1994). (Reproduced by permission of Elsevier Science Publishers B.V.)

height of the stratified layer, h_2, and the concentrations in the two layers, c_1 and c_2. The flammable mass is found by knowing the concentrations and layer heights.

An example of the adequacy of the model predictions is shown in Figure 5.23. Since the LFL for methane is 5.0%, the sonic release predictions are just below the LFL, whereas some data are above the LFL. Taking the lower limit as 0.9LFL would give a nearly correct result, though, for the sonic release case. The model does better for the subsonic release, predicting that the volume above 1.5 m would be above the LFL.

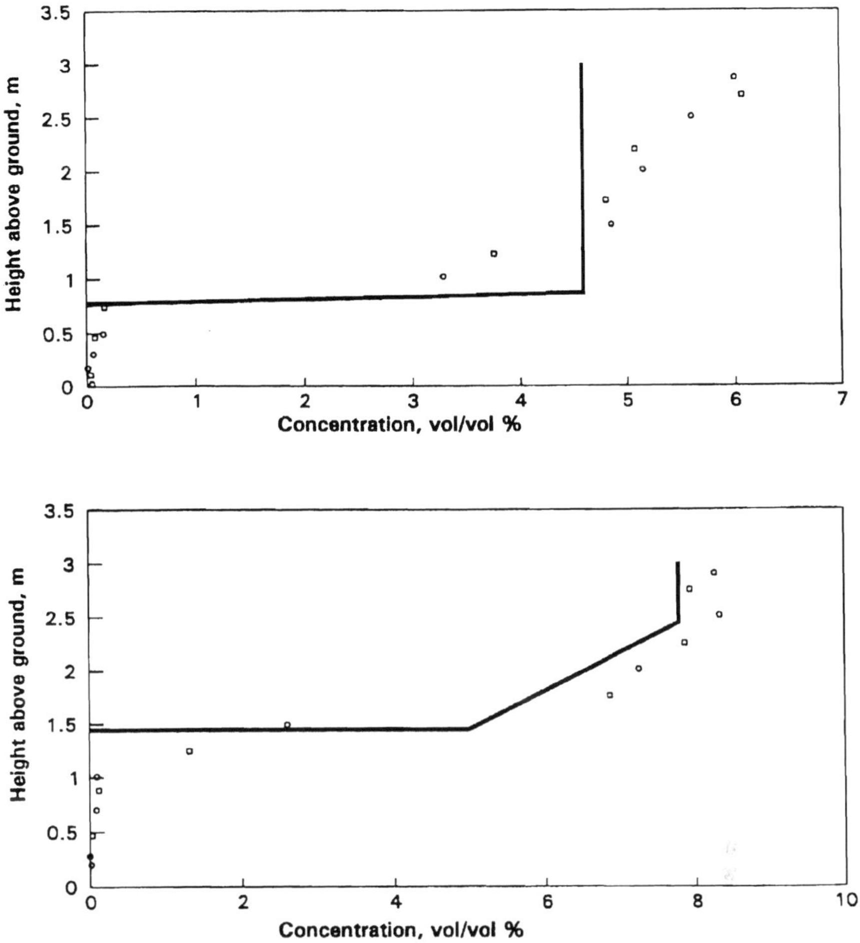

Figure 5.23. Comparison of predictions of the model for gas injection into an enclosed room with experimental data (Cleaver et al., 1994). (Reproduced by permission of Elsevier Science Publishers B.V.)

6

OVERVIEW OF RELATED COMPUTER PROGRAMS

It ain't what you don't know that gets you;
it's what you know that ain't so.
—Texas Expression

Any list of computer programs that calculate discharge rates, vapor cloud dispersion, and flammable mass is ever changing and completeness is nearly impossible to achieve. Dispersion modeling principles and models are treated in other CCPS publications (CCPS, *Guidelines for Use of Vapor Cloud Dispersion Models,* 2nd ed., 1996; CCPS, *Understanding Atmospheric Dispersion of Accidental Releases*, 1995). In this chapter we provide a sample of available computer programs by listing those treated in major model validation and comparison studies. Such studies are subject to rapid obsolescence since updated versions are issued approximately annually of all models that are under active maintenance. Furthermore, more recent experimental data are not included in earlier model evaluation studies, and model performance varies with the experimental tests used in the comparison.

Models are cited from the following validation projects:

- American Petroleum Institute (API) model evaluation study (API, 1992) and Hanna et al. (1993)
- SMEDIS Project (Scientific Method for Evaluation of Dispersion Models) planning documents and progress reports (SMEDIS, 1997)

The SMEDIS project is planned for a 3-year span with the objectives of developing a protocol for model evaluation and using this protocol with special emphasis on the more difficult tasks of modeling the effects of aerosols, complex terrain, and obstacles. As part of the protocol development, the project has assembled a literature file of available experimental tests, and will evaluate and select the best quality data sets for use. Models will be evaluated theoretically as well as by comparison of model predictions with data. A thorough technical review will include comparing actual coding with model documentation for congruence.

Table 6.1 lists the models evaluated in these projects and brief descriptions of the models and their treatment of flammable mass. The latest versions of these models are listed, although earlier versions were evaluated in the API project. The categories of models follows the SMEDIS project approach using the following abbreviations:

SC = Screening Tool
GA = Gaussian model
IN = Integral model
SLT = Shallow layer with terrain effects
CFD = Computational Fluid Dynamics

The convention used in Table 6.1 is

✔ = feature is in model
(✔) = feature is available as a separate program
— = feature is not in model

TABLE 6.1
Properties of Dispersion Models

Model name	Britter-McQuaid Workbook	VDI3783 Part 2	AFTOX	ADAM	ALOHA	CHARM	INPUFF	AIRTOX	CANARY
GENERAL INFORMATION									
Version number	1	4.0	3.1[1]	2.1	5.2	6.1	2.0	?	3.0 [1]
Date	1988	1990	1988	1987	1996	1991	1988	1988	1997
Developer	R E Britter J McQuaid of HSE	Met. Inst. U. Hamburg Germany	US Air Force, Kunkel	US Air Force, Raj & Morris	NOAA	Radian Corp.	EPA, Peterson & Lavadas	ENSR Consulting, Mills	Quest Consultants
Model Type	ST	ST	GA	GA	GA	GA	GA	GA	IN
MODEL CAPABILITIES									
Jet submodel									
Momentum jet source	—	(✔)	—	—	—	—	—	—	✔
Ground level source	—	✔	✔	—	—	—	—	—	✔
Elevated source	—	—		—	—	—	—	—	✔

Model name	Britter-McQuaid Workbook	VDI3783 Part 2	AFTOX	ADAM	ALOHA	CHARM	INPUFF	AIRTOX	CANARY
Liquid pool evaporation									
Vaporizing pool source	—	(✔)	—	—	—	—	—	—	—
Time variation of release rate									
Instantaneous	✔	✔	✔		✔	✔	✔	✔	✔
Continuous	✔	✔	✔	—	✔	✔	✔	✔	✔
Finite duration continuous	✔[1]	✔	—	—	✔	✔	✔	✔	✔
Two-phase releases	—	✔	—	—	✔	✔	✔	—	✔
Output	✔	✔	✔		✔	✔	✔	✔	✔
Concentration along centerline	✔	✔[1]	✔	—	✔	(✔)	✔	✔	(✔)
Concentration at arbitrary points	✔ [2]	✔	✔	—	✔	(✔)	✔	✔	(✔)
Flammable mass	(✔) [2]	✔	(✔) [2]	—	(✔)	(✔)	(✔)	✔	(✔)

TABLE 6.1
Properties of Dispersion Models (continued)

Model name	DEGADIS	DRIFT	GASTAR	HGSystem	HAGAR	PHAST, SAFETI, UDM[1]	OHRAT's dispersion models	WHAZAN
GENERAL INFORMATION								
Version number	2.1	2.23	2.96 [1]	3	4	5.2	1.3.1	2.1
Date	1988	1991	1996	1994	Dec 1996	May 1997	Jun 96	May 93
Developer	U. of Arkansas, Havens & Spicer	AEA Technology UK	CERC, UK	SHELL, UK	British Gas, UK	DNV, UK and USA	DNV, UK and USA	DNV, UK and USA
Model Type	IN	IN	IN	IN	IN	IN	IN	IN
MODEL CAPABILITIES								
Jet submodel								
Momentumjet source	(✔)	(✔) [1]	✔	✔	(✔) [1]	✔	(✔)	(✔)
Ground level source	✔		✔	✔	—	✔	—	—
Elevated source	✔		✔	(✔)	✔	✔	✔	✔ [1]

Model name	DEGADIS	DRIFT	GASTAR	HGSystem	HAGAR	PHAST, SAFETI, UDM[1]	OHRAT's dispersion models	WHAZAN
Liquid pool evaporation								
Vaporizing pool source	✔	(✔) [2]	✔	✔	(✔)	✔	(✔)	(✔) [2]
Type of release rate								
Instantaneous	✔		✔	✔	✔	✔	✔	✔
Continuous	✔	—	✔	✔	✔	✔	✔	✔
Finite duration continuous	✔	—	✔	✔	✔	✔	✔	✔
Two-phase releases	✔	—	✔	✔	✔	✔	✔	✔
Output								
Concentration along centreline	✔	✔	✔	✔	✔ [2]	✔ [2]	✔ [1]	✔
Concentration at arbitrary points	—	—	(✔)	—	(✔) [2]	✔	—	—
Flammable mass	✔	—	—	—	(✔) [2]	✔	—	—

TABLE 6.1
Properties of Dispersion Models (continued)

Model name	TRACE	ESCAPE	AERCLOUD	EOLE	GReaT	SLAB	SLUMP
GENERAL INFORMATION							
Version number	2.5.4	6.10	1.0	3.0	2.0	None	2B
Date	1989	Nov 96	15-04-93	Dec 1996	Jan 94	Jun 1990	1989
Developer	SAFER, Inc.	Finnish Met. Institute, Finland	Finnish Met. Institute, Finland	Gaz de France, France	Risø Nat. Lab, System Analysis Denmark	D.L. Ermak, Lawrence Livermore Nat Lab, USA	WS Atkins Safety & Reliability UK
Model Type	IN	IN	IN	IN	IN	IN[1]	IN
MODEL CAPABILITIES							
Jet submodel							
Momentumjet source	✔	✔	—	—	✔	✔	—
Ground level source	✔	✔	—	—	✔	✔	—
Elevated source	✔	—	—	—	✔	—	—

Model name	**TRACE**	**ESCAPE**	**AERCLOUD**	**EOLE**	**GReAT**	**SLAB**	**SLUMP**
Liquid pool evaporation							
Vaporizing pool source	(✔)	✔	—	(✔)	—	✔ [2]	—
Type of release rate							
Instantaneous	✔	✔	✔	✔	—	—	✔
Continuous	✔	—	—	✔	—	✔	—
Finite duration continuous	✔	—	—	✔	—	✔	—
Two-phase releases	✔	✔	✔	—		✔	—
Output							
Concentration along centerline	✔	✔	✔	✔	[1]	✔	✔
Concentration at arbitrary points	✔ [1]	(✔)	—	(✔)	—	✔	✔
Flammable mass	—	(✔)	—	✔	—[2]	✔	✔

TABLE 6.1
Properties of Dispersion Models (continued)

Model name	DISPLAY-1	DISPLAY-2	SLAM	TWODEE	CFX	MERCURE	COBRA	ADREA-HF	STAR-CD
GENERAL INFORMATION									
Version number	1.0	1.0	1.0 [1]	2.11	4.1	3.2	4z06	2nd version	2.300
Date	July 1996	Nov 1996	Jun 1996	Feb 1997	1996	12/96	Sep 1995	Sep 1996	Jun 1996
Developer	JRC, Italy	JRC, Italy and NCSR Demokritos Greece	Risø National Lab., Denmark	HSL (HSE), UK	AEA Tech-nology, UK		Mantis Numerics Ltd, Leeds	necessary, Greece in collaboration with JRC ISPRA, Italy	Computa-tional Dynamics Ltd, UK
Model Type	SLT	SLT	SLT	SLT	3D [1]	3D	3D	3D	3D
MODEL CAPABILITIES									
Jet submodel									
Momentum jet source	(✔)	✔	—	—	✔	✔	✔	✔	(✔)
Ground level source	✔	✔	—	—	✔	✔	✔	✔	
Elevated source	—	✔	—	—	✔	✔ [1]	✔	✔	

Model name	DISPLAY-1	DISPLAY-2	SLAM	TWODEE	CFX	MERCURE	COBRA	ADREA-HF	STAR-CD
Liquid pool evaporation									
Vaporizing pool source	(✔)	(✔)	—	—	(✔)[2]	✔	✔	—	(✔)
Type of release rate	✔	✔	✔	✔	✔	✔	✔	✔	✔
Instantaneous	✔	✔	✔	✔	✔	✔	✔	✔	✔
Continuous	✔	✔	—	✔	✔	✔	✔	✔	✔
Finite duration continuous	✔	✔	—	✔		✔	✔	✔	✔
Two-phase releases	✔	—	—	—	— [3]	✔	—	✔	✔
Output									
Concentration along centerline	✔	✔	—	✔ [1]	(✔)	✔	✔	(✔)	✔ [1]
Concentration at arbitrary points	✔	✔	—	✔	(✔)	✔	✔	(✔)	✔ [1]
Flammable mass	(✔)	(✔)	—	—	(✔)	✔ [2]	✔	(✔)	(✔) [1]

Notes on Models

Britter-McQuaid Workbook

[1] The yes answer here must be qualified. The finite duration continuous releases are constrained to fit either instantaneous or continuous releases.
[2] The cloud is assumed to take a 'box' shape throughout and calculations arising from this assumption can be performed

VDI 3783, Part 2 (German Equivalent of ASTM)

[1] In principle centerline concentrations can be found.

AIRTOX

[1] Lotus 1-2-3 spreadsheet based.
[2] See Appendices I, J, K.

CANARY

[1] Derived from SLAB. Replaces FOCUS Version 2.1.

DRIFT

[1] The external jet model is EJECT 1.01
[2] The external pool evaporation model is GASP 3.01

GASTAR

[1] Validation documentation are available for an earlier version (Hanna's study).

HAGAR

[1] HAGAR cannot deal with a jet directly into the wind, but downwind jets and crosswind jets at any angle are allowed
[2] These are available at all times for an instantaneous releases. For continuous releases, a release which is equivalent to the worst case dispersion distance is found and modeled. For slowly decaying releases, such as a pool of liquefied gas remaining in a diked region, concentrations are also available for the time during which the release is slowly decaying.

PHAST, SAFETI, UDM

[1] Table entries refer to PHAST and SAFETI, respectively.
[2] Pseudo-component (averaged) properties are adopted for a multicomponent mixture.

OHRAT's dispersion models

[1] Pseudo-component (averaged) properties are adopted for a multicomponent mixture.

WHAZAN

[1] Elevated release only as free horizontal jet above ground.
[2] Pool evapaoration model requires user-specified rainout fraction.

GReAT

[1] Treats multicomponents only if defined as a pseudo-single substance.
[2] Calculates flammable mass, but with minor effort.

SLAB

[1] The following is cited from User's manual of SLAB:
"The SLAB model is built upon a theoretical framework that starts with averaged forms of conservation equations for mass, momentum, energy and species." It does not treat terrain effects, although claiming to be based on shallow layer modeling equations.
[2] Pool evaporation treated in EFFECTS 2.1, not in SLAB code.

TRACE

[1] Uses top hat coupled to Gaussian model

SLAM

[1] The model is a research tool right now: there is not yet a user interface or manual. Developers intend to extend the model for time dependent releases (jets) and to include heat transfer.

TWODEE

[1] Some of these variables have little meaning for shallow layer models where the plume follows the terrain. At any time the model will give the shape and concentration distribution in the cloud.

CFX

[1] The code is machine specific. It can be run on mainframes and parallel machines in different versions. Most users use workstations.
[2] Requires separate pool model to supply boundary conditions for CFX.
[3] Model requires extension for aerosol releases. Does not have built-in treatment of moisture in the atmosphere.

MERCURE

[1] Requires pre-processing to develop vertical wind speed profile.

[2] Requires post-processing to calculate flammable mass.

STAR-CD

[1] Post-processing is carried out using PROSTAR. This enables the following to be obtained very easily:

- concentrations in % vol or volume ppm, etc. (a user subroutine could also be used instead of doing this as a post-processing operation);
- mass of gas between two concentration levels;
- concentration profiles along arbitrary lines;
- mass flux across arbitrary planes;
- cloud dimensions in any direction, e.g. width, height, hazard range;

cloud arrival times.

7

WORKED EXAMPLES

> The future is merely a reenactment of the past. Industrial accidents for the most part are not new occurrences. Their cause is rooted deep in the past minor incidents, where damage was insignificant and workers or bystanders were not injured.
>
> —Kletz, 1993

Finding the flammable mass for a vapor cloud is primarily an exercise in using a dispersion model to generate predictions of the vapor cloud concentration profiles that vary in space and time. Some dispersion models have built in routines to integrate the resulting vapor cloud concentration profiles between the LFL and UFL to obtain the flammable mass. For purposes of illustration some examples below were calculated with the PHAST model. This is not intended to be an endorsement, merely a convenience. An alternative solution is to match the results from a dispersion model to a Gaussian vapor cloud and then use the analytic formulas given in Appendices I, J, and K to calculate the flammable mass. Examples are given below for the three types of vapor clouds discussed in Chapter 5: unconfined, partially confined, and confined.

After finding the flammable mass in a vapor cloud, the next problem is to estimate the flammable mass in that portion of the cloud that occurs in congested and confined areas. Some alternative methods are listed in Table 7.1 in order of increasing realism and complexity. Methods 1, 2 and 3 in Table 7.1 are illustrated in Section 7.3. Method 4 is illustrated in Section 7.4. Methods 3 and 4 adjust for the variable density of the cloud as a function of downwind distance.

TABLE 7.1
Alternative Methods for Finding Flammable Mass in Congested Areas

1. Disregard concentration gradients. Assume concentration is uniform in each congested area at the stoichiometric concentration. Use a simplification that the heat of combustion per unit volume is 3.5 MJ/m^3. (Section 4.4) This is suggested for the original Multi-Energy method.
2. Find the fraction of the total footprint area of a vapor cloud that is in a congested area and multiply this fraction by the overall flammable mass found by the dispersion model for the cloud.
3. Use the density gradient reported by a dispersion model. Find the average density of the cloud in each congested area. Assume concentration is uniform in each congested area. Multiply the cloud volume in each congested area by the corresponding average density.
4. Divide the flammable cloud into sections perpendicular to the x-axis (along-wind direction) using equal time increments and the cloud velocity profile. Integrate the flammable mass in each such section. Estimate the fraction of each section that is within each congested area. Add the flammable mass for the portion of each section in each congested area.

7.1. Example 10, Unconfined Vapor Cloud—Vapor and Liquid Propane Releases

As an example of an unconfined vapor and liquid release, consider a 30,000-gallon storage tank of propane at 25 C (298 K). Consider two cases: (1) a line break from the vapor space and (2) a line break from the liquid space. The mass discharge rate is found by Eqs. (4.2) for a subcooled liquid release and Eq. (4.6) or (4.8) for a vapor release. Interestingly, the same mass rate is calculated for a 4-in. line discharging vapor and a 1½-in. line discharging liquid (see Examples 2 and 3 in Chapter 4). Dispersion will be greater for a vapor release than for a liquid release because the expansion velocity found by Eq. (N.4) in Appendix N will be substantially higher as listed in Table 7.2. As the vapor cools by expansion, some condensation occurs, and similarly, the liquid flashes partly to vapor, so both discharges become two-phase [see Eq. (4.14)]. The discharge coefficient is 0.61 for liquid discharge, and increases with vapor quality to 0.88 for the two-phase release from the vapor space using formulas by Bragg (1960) or Jobson (1955) in Appendix N.

The API "Guide for Pressure-Relieving and Depressuring Systems" may be used to estimate the size of a vertical jet from a flare stack. For the more general case of releases at any angle, other jet models are needed. The flammable mass and plant area covered by a flammable propane vapor cloud from these discharges as predicted by a model are summarized in Table 7.3. These results depend on the direction of discharge.

TABLE 7.2
Discharge Conditions for Propane from the Vapor and Liquid Space of a Storage Tank

Discharge Type	Release Direction	Mass Fract. of Liquid	Drop Size, (m)	Discharge Coefficient	Expansion Velocity (m/s)
Vapor space	Up	0.0538	0.1	0.88	442
Vapor space	Horizontal	0.0538	0.1	0.88	442
Liquid space	Down	0.662	7.0	0.61	211
Liquid space	Horizontal	0.662	7.0	0.61	211

A *vapor* discharge directed upward provides ample opportunity for dilution before it slumps to ground level. This case gives the smallest flammable mass at ground level (zero) because the cloud concentrations at ground level are below the LFL. However, the cloud dimensions to 0.9 LFL and 0.5 LFL at ground level are finite, so pockets of flammable material may exist out to 54 m according to Table 7.3. The same discharge directed horizontally produces a flammable mass of 41.6 kg, and the ground-level 0.5LFL contour extends near twice as far.

A *liquid* discharge directed downward loses some of the jet momentum and consequently disperses less well than a horizontal release of liquid. A downward liquid release produces by far the largest flammable mass in Table 7.3. A horizontal liquid release produces the second-largest flammable mass.

Ambient temperature 25 C; saturated liquid at 9.51 bar(a) pressure, urban or plant conditions (roughness length z_0 = 1 m); release elevation = 2 m for liquid, 4 m for vapor; liquid head = 1.5 m; 4-in, vapor line; 1.5-in. liquid line; initial discharge rate w =19.93 kg/s (2636 lb./min) in both cases.

TABLE 7.3
Dispersion and Flammable Mass Predictions for Propane Releases of Table 7.2

Discharge Type	Release Direction	Flam. Mass (kg)	Distance, x to 0.9 LFL (m)	Elevation, z to 0.9 LFL (m)	Distance x to 0.5 LFL (m)	Elevation, z to 0.5 LFL (m)
Vapor space	Up	None at ground level[a]	3.8	40.6	10.0	53.6
Vapor space	Horizontal	41.6	60	3.0	102	0
Liquid space	Down	2622	98	0	134	0
Liquid space	Horizontal	863	92	0	153	0

[a]Plume bends over in the wind, but dilutes below the LFL before reaching the ground.

7.2. Example 11, Unconfined Vapor Cloud—Effect of Wind Speed

The effect of wind speed on the flammable mass in a vapor cloud is illustrated by the rupture of a 4-in. line from the bottom of a partially full 30.5 ft (10 m) diameter sphere of vinyl chloride monomer (VCM). In cases when the pipe diameter is small in comparison with the vessel diameter, the discharge rate is nearly constant or slowly decreasing, so it is justified to assume a steady-state rate. The conditions and flammable mass calculations by a model are summarized in Table 7.4 for several wind speeds. Flammable mass from Table 7.4 is plotted in Figure 7.1. A plan view plot of the unconfined vapor cloud is shown in Figure 7.2 for a 5 m/s wind speed.

Ambient temperature = 25 C, atmospheric stability = D, tank diameter = 10 m, liquid head = 6 m, line length = 10 m, release height = 1 m. Discharge rate = 59.1 kg/s, Sauter mean droplet diameter = 43.5 μm, expansion velocity = 101 m/s. As can be seen, the flammable mass is a sensitive inverse function of wind speed.

7.3. Example 12, Partially Confined Vapor Cloud Explosion—Vinyl Chloride Monomer Release

The VCM release example is continued here to further illustrate the procedure for a partially confined explosion analysis.

Step 1 in the procedure in Section 4.8.4, is to identify PES areas. At the same time assign the degree of confinement and of congestion for PES areas, Steps 7 and 8 of the procedure. This is illustrated in Figure 7.3 using Baker–Strehlow classifications.

Step 2 in the procedure is to plot the cloud footprint to the LFL, and then to overlay this footprint on the plot plan. For example, in Figure 7.4 the footprint

TABLE 7.4
Conditions and Flammable Mass for Example 11 4-in. Line Break from a VCM Storage Sphere

Wind speed, m/s	1.5	2.5	5.0	7.5	10
Rainout, mass %	21	21	18	16	16
Pool diameter, m	19	19	17.3	16	16.2
Vapor rate, kg/s	51	51	52	53	53
Max. x to LFL, m	165	172	175	159	152
Flammable mass, kg	1450	983	578	298	206

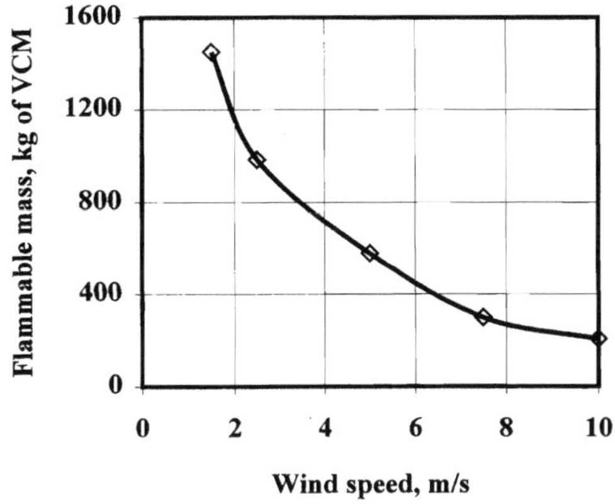

Figure 7.1. Effect of wind speed on flammable mass of VCM vapor cloud.

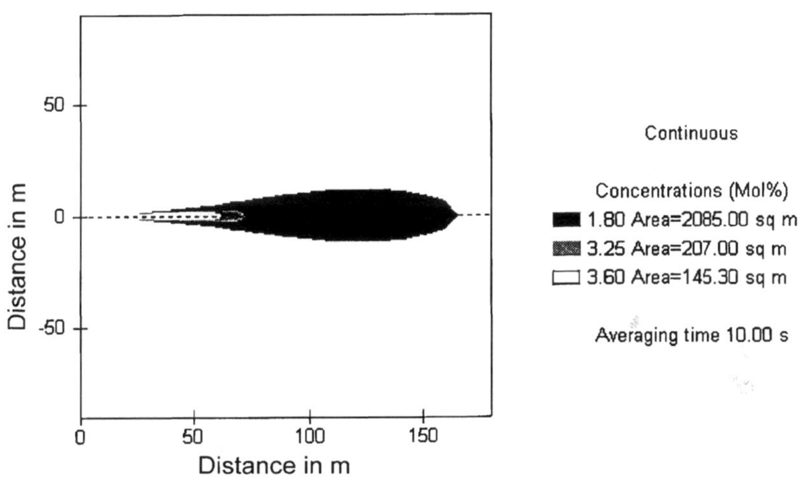

Figure 7.2. Cloud footprint for VCM pipe break, 5 m/s wind speed, D stability.

for the LFL contour (3.60 % VCM) from Figure 7.2 is shown to overlay one congested area, classified as 2D,H (two-dimensional degree of confinement, high degree of congestion).

Step 3 in the procedure is to estimate the cloud height and confinement height over each congested area. Figure 7.5 plots effective cloud height as predicted by the dispersion model for three wind speeds. For a 5 m/s wind speed, the cloud height grows between about 4 m and 7 m over the 2D,H congested area. Taking the effective deck height of the congested area to be 5 m, then the cloud

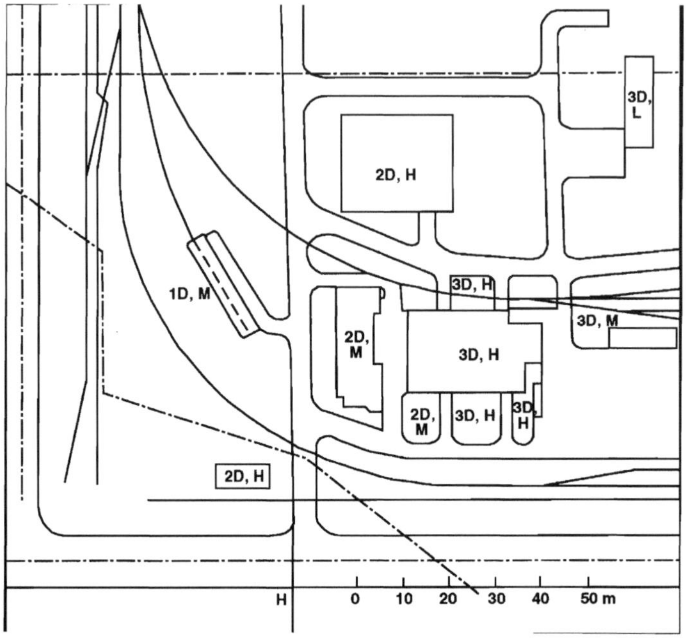

Figure 7.3. Assigning Baker–Strehlow confinement, congestion designations to PES areas on a typical plot plan.

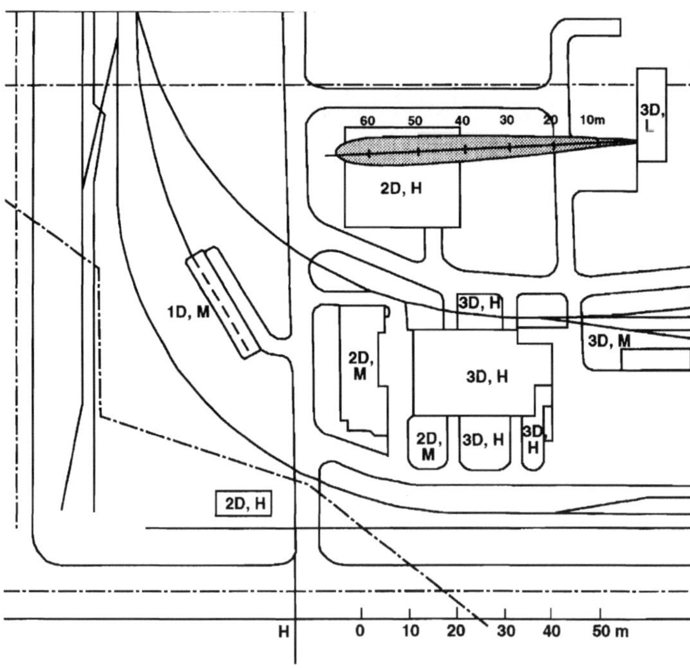

Figure 7.4. Overlay of vapor cloud footprint from 4-in. VCM pipe break on plot plan

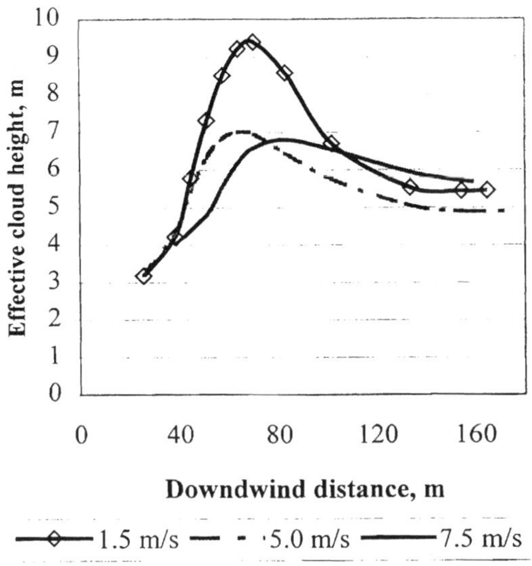

Figure 7.5. Effect of wind speed on cloud height for a 4-in. VCM line break.

would be constrained by the deck, so the classification of 2D flame expansion is justified and the cloud height will be held to 5 m.

For step 4 in the procedure, estimating the flammable mass in each PES, we make use of Methods 1, 2 and 3 from Table 7.1. In step 3 we found that the vapor cloud will be under the deck height in the 2D,H congested area. If not, treat two cloud portions, one above and one under the deck. The area of the cloud that overlaps the congested area is found by integrating over Figure 7.4 giving 87.2 m^2 or 60% of the total footprint area. The volume is $5 \times 87.2 = 436$ m^3. Since the nose of the cloud overlaps the congested area only slightly, it is included in the integration.

According to Method 1 of Table 7.1 only the volume of the cloud in the congested area is needed to find the energy in an explosion (flammable volume times approximate volumetric heat of combustion):

$$V_f H_{cv} = (436)\ (3.5\ \text{E6}) = 1526\text{E6 Joules}$$

Since the heat of combustion of VCM is 18.53 MJ/kg, the flammable mass predicted using Method 1 is

$$m_f = 1526\ \text{E6}/\ 18.53\ \text{E6} = 82.4\ \text{kg of VCM}$$

Using Method 2 of Table 7.1 the flammable mass for the total cloud as listed in Table 7.4 for a 5.0 m/s wind speed is 578 kg. Multiplying by the ratio of the vapor cloud area in the congested area to the total cloud area estimates the flammable mass in the congested area as

$$m_f = (0.6)(578) = 346.8 \text{ kg of VCM}$$

This estimate tends to be high since it weights equally all areas in the cloud, whereas because of concentration gradients the nose of the cloud has a lower density of flammable mass than does the area nearer to the source.

With Method 3 of Table 7.1, use information from the dispersion model that average cloud density varies from 1.49 kg/m^3 at 43 m from the source to 1.41 kg/m^3 at 66 m from the source. The average density over the congested area is 1.45 kg/m^3. Subtracting the ambient air density of 1.173, so the mass of VCM in the cloud overlapping the congested area is

$$m_f = (1.451 - 1.173)\ 436 = 121 \text{ kg of VCM}$$

The Method 3 estimate will be used in remaining calculations for this example case.

Step 5 of the blast analysis method has been satisfied by finding the heat of combustion of VCM listed above.

Step 6 is to select the reactivity of the fuel, which, according to Table 4.1 lists VCM as medium reactivity based on the laminar flame speed. For the Multi-energy method assume ignition energy is low.

Steps 7 and 8 were done along with Step 1. Step 9 requires finding the maximum overpressure parameter. For the Baker–Strehlow method, with 2D confinement, medium fuel reactivity, and high obstacle density, Table 4.10 gives a value of 1.24 as the maximum overpressure parameter expressed as turbulent flame speed. For the multi-energy method, Table 4.11A gives the maximum overpressure parameter value of 5–7 for 2D confinement, high obstacle density, and low ignition energy. For Step 10, calculate overpressure and impulse vs. distance curves from Figure 4.55, and use these to predict blast loads on specific buildings. The energy scaling factor for the example explosion is

$$ES = \left(\frac{m_f H_c}{P_0}\right)^{1/3} = \left(\frac{121 * 1.853 \times 10^7}{101325}\right)^{1/3} = 28.1 \text{ (m)}$$

Values are read from the dimensionless plots of overpressure and impulse as a function of the dimensionless distance:

$$\overline{R} = \frac{R}{ES}$$

Table 7.5 lists the dimensionless coordinate points read for the Baker–Strehlow method and then translated to dimensional values. The process is repeated in Table 7.6 using the multi-energy method. The dimensionless impulse is plotted for the Baker–Strehlow method as

$$\bar{I} = \frac{I_e a_0}{P_0 ES}$$

This is solved for I_e. The dimensionless pulse duration for the multi-energy plots is defined as

$$\bar{t}_+ = \frac{t_+ a_0}{ES}$$

The pressure pulse is taken to be a triangle with peak overpressure ΔP_e and duration t_+, so the impulse is given by

$$I_e = \tfrac{1}{2} \Delta P_e t_+$$

For this example the multi-energy method predicts slightly higher impulses, while the Baker–Strehlow method predicts higher peak overpressures.

TABLE 7.5
Blast Wave Predictions by the Baker–Strehlow Method for Example 12, Vapor Cloud Partially Overlapping a Single Congested Area

$\bar{R}$ (—)	0.25	0.5	1.0	2.0	3.67	5	10
R (m)	7.0	14	28	56	103	140	280
$\Delta P_e/P_0$	3.1	1.01	0.32	0.19	0.080	0.066	0.038
$\dfrac{I_e a_0}{P_0(E/P_0)^{1/3}}$	0.118	0.0596	0.0339	0.0187	0.010	0.0078	0.00396
t_+ (ms)	6.0	9.5	17	16	20	19	17
I_e (Pa·s)	940	480	270	150	80	62	32

TABLE 7.6
Blast Wave Prediction for Example 12 by Multi-Energy Method
Overpressure Parameter = 6

$\Delta P_e/P_0$	0.50	0.50	0.30	0.136	0.068	0.0467	0.0219
$\bar{R}$ (—)	0.25	0.5	1.0	2.0	3.67	5	10
R (m)	7.0	14	28	56	103	140	280
$\bar{t}_+$ (—)	0.5	0.36	0.375	0.405	0.43	0.45	0.47
t_+ (ms)	39	28	30	32	34	36	37
I_e (Pa·s)	998	720	450	220	117	84	41

7.4. Example 13, Partially Confined Vapor Cloud Explosion—Total Petroleum LaMede Refinery Explosion, November 1992

A vapor release and explosion occurred in November 1992 at Total Petroleum's LaMede Refinery which is just inland from the Mediterranean Sea in France. There were no special operations in progress at the time of the incident, and the operating personnel had not noticed any abnormal conditions. An inquiry established that a leak occurred on an 8-in. sidestream of the deethanizer column in the light-ends plant which recovers propane and butane from the catalytic cracker product material. The column operated at 10 bar and 40 C. The origin of the leak was evidently internal pitting corrosion that resulted in a hole size of 25 $cm.^2$ (equivalent to a hole slightly larger than 2 in. diameter). An investigating team estimated that 10 tonnes of material was released over about a 7–10-min period.

The estimated inventories of flammable material available to participate in the original release and explosion as well as in the subsequent fires are listed in Table 7.7.

An explosion occurred centered in the gas plant. The investigators estimated that about 5 tonnes of hydrocarbon took part in the explosion. A plot plan in Figure 7.6 shows the nearby units and sketches damage contours. The damage is summarized in Table 7.8.

Based on these assessments, the investigators assigned the damage radii in Table 7.9. These are plotted in Figure 7.6. The center of the damage circles is taken at an estimated center of the explosion.

Estimation of Flammable Mass Using Modeling

The LaMede Refinery example can be taken to illustrate a predictive study (assuming the accident information is not known) as well as an accident analysis study. The first task is to calculate the mass discharged. For a puncture in a distillation column sidestream the initial discharge is likely to be a liquid. However,

TABLE 7.7
Estimated Initial Inventories Before LaMede Refinery Explosion

Deethanizer, depropanizer, debutanizer	100 tonne
Main fractionator overhead receiver	30 tonne
Gasoline fractionator plus associated equipment	50 tonne
Pipework, etc.	20 tonne
Total	200 tonne

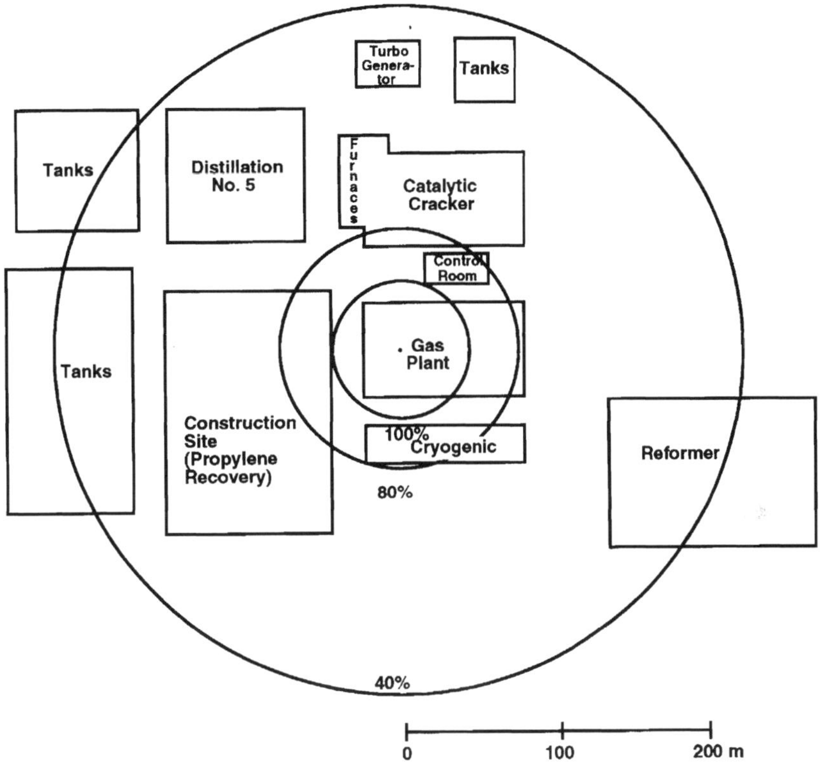

Figure 7.6. LaMede Refinery Plot Plan and Explosion Damage Contours Estimation of Flammable Mass Using Modeling.

TABLE 7.8
Damage Observations from LaMede Refinery Explosion

Gas plant	Damaged beyond economic repair.
Control room	Completely destroyed.
Cryogenic unit	Damaged beyond economic repair.
Turbo alternator	Building and plant severely damaged.
Reformer	Medium scale damage.
Distillation Unit No. 5	External damage, but repairable.
Construction site	Badly damaged. Crane toppled.
Catalytic Cracker	Damage to supports and internals.

TABLE 7.9
Overall Damage Radii for LaMede Refinery Explosion

100% damage radius	40 m
80% damage radius	70 m

there is limited liquid inventory on the trays in a distillation column, so the release is likely to change from a liquid to a vapor discharge. A detailed accident analysis would use a simulator for a distillation column that is capable of calculating liquid and vapor holdup and pressure decay during the release. For this example, we use simpler assumptions.

Another complication is that the composition of material in a distillation column varies. Fortunately, for flammable mass calculations, it is often accurate enough to use a single-component surrogate material for a multicomponent mixture. The example calculations below show how to use a single-component.

The composition in the deethanizer column could range between ethane at the condenser to butanes at the reboiler. Olefins are present in catalytic-cracked material. For the components in Table 7.10 the vapor pressure at 313 K, $P_{s,}$ is listed, along with the normal boiling point (NBP), densities at 313 K and flammable limits.

At the column pressure of 10 bar(a), as pure components, only butanes will be a liquid. Of these we choose, rather arbitrarily, ethane to represent vapor discharge and1-butene to represent liquid discharge. The formulas in Section 4.1.1 give the discharge rates in Table 7.11. For gas discharge, these values are obtained using ethane ($P_o = 10$ bar, $\rho_{vo} = 12.34$ kg/m^3). The expansion velocities are found by Eq. (N.4), showing that vapor discharge generates much more jet momentum mixing than liquid discharge.

TABLE 7.10
Physical Properties of Components in the Deethanizer Column

Component	NBP, K	P_s at 313 K, bar(a)	Liquid density, kg/m^3	Vapor density at 1 atm., kg/m^3	LFL, mole %	UFL, mole %
Ethane	184.5	48.8	200.9	1.178	3.0	15.3
Propylene	225.4	16.5	476.3	1.659	2.0	11.0
Propane	231.1	13.7	468.5	1.741	2.1	9.5
Iso-butane	261.4	5.34	533.1	2.315	1.8	8.4
1-Butene	266.9	4.55	567.6	2.236	1.6	9.3

TABLE 7.11
Liquid and Vapor Discharge Rates for LaMede Refinery Incident

Component	G^*	G, kg/m²s	C_D	w, kg/s	u_{exp},m/s
1-Butene (liquid)	0.642	15,300	0.61	38.3	107
Ethane (vapor)	0.646	2,267	0.88	5.00	633

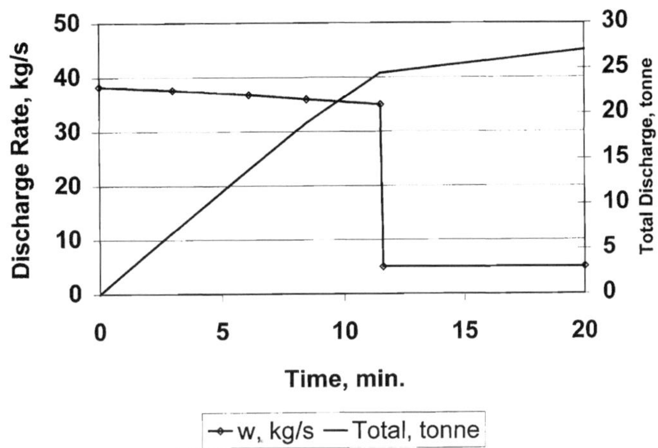

Figure 7.7. Predicted discharge rate for the LaMede Refinery incident.

Time-varying blowdown calculations are done using a dispersion program to produce the discharge predictions plotted in Figure 7.7. The discharge is assumed to be liquid for 11 min, followed by vapor discharge. At 11 min after the break, the predicted mass released is 24.5 tonnes of hydrocarbon. As a quality assurance check, 24.5 tonnes provides enough explosion energy that blast predictions are consistent with the observed damage.

To model explosions in partially confined, congested areas, again follow the procedure listed in Section 5.3.1. Following Step 1 we identify areas of congestion and confinement likely to promote flame acceleration to an explosion, or possible explosion sites (PES). The following plants will be considered PESs:

- Gas plant
- Catalytic cracking plant
- Turbo-generator plant
- Control room
- Some tanks

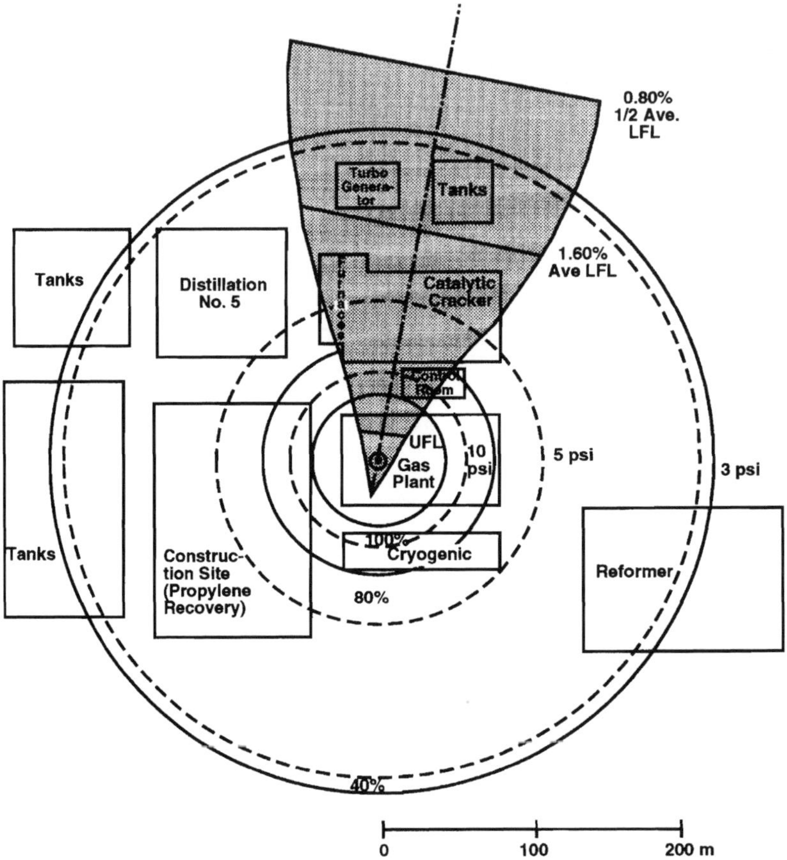

Figure 7.8. Vapor cloud profile predicted for LaMede Refinery release.

TABLE 7.12
Dispersion Modeling Conditions Used for LaMede Refinery Incident

Wind Speed, Stability	3.4 m/s, F
Ambient Temperature	298 K
Relative Humidity	50 %
Atmospheric Pressure	1.01325E5 Pa
Release Height	0.76 m (2.5 ft)
Release Direction	Down
Roughness Length, z_0	1.0 m

TABLE 7.13
Vapor Cloud Profiles Predicted by a Dispersion Model

x,m	Centerline Conc., mole %	Effective cloud width, m	Effective cloud height, m	Significance
6.3	40.5	1.7	1.5	
41	9.3	10	7.1	UFL, maximum height
100	4.30	86	5.0	
167	1.80	149	6.2	Average LFL
180	1.60	158	6.5	LFL of 1-Butene
263	0.80	187	9.3	0.5 LFL of 1-Butene

Step 2 is to use a dispersion model and plot the predicted plan view footprint of the cloud between the LFL and UFL. Dispersion modeling is done using the conditions in Table 7.12. The plume profile predicted by a dispersion model is plotted in Figure 7.8.

Following Step 3 we compare the predicted cloud height with the height of each PES. Selected points from the dispersion model predictions are shown in Table 7.13. The height of pipe racks, platforms, fin-fan coolers and the like is taken to be 6 m, so the vapor cloud falls under this height until it dilutes to the LFL of 1.80%.

Step 4 is to estimate the flammable mass in each PES. For this example, make use of Method 4 in Table 7.1. Method 4 takes advantage of the fact that in each lateral slice of the cloud passing a point for 1 sec, with a discharge rate of w =46.5 kg/s there is very nearly 46.5 kg of flammable material. That is, there is minimal diffusion in the along- wind, x, direction, so the mass in a vertical slice perpendicular to the wind equals the discharge rate multiplied by the transit time of the slice. The arrival time of the cloud front can be taken from the results of the dispersion modeling. The flammable mass is found by the following formula:

$$m_{\mathrm{f}} = w \sum_i \left(t_{\mathrm{leave}} - t_{\mathrm{start}}\right)_i \frac{W_{\mathrm{PES}i}}{W_{\mathrm{cls}}} \tag{7.1}$$

where w is the discharge rate. The second term is the time it takes for the cloud to pass through the PES. The third term is the width fraction, or the width of the PES that is in the cloud divided by the width of the cloud (averaged over the downwind distance). This calculation is detailed in Table 7.14.

The integration of flammable mass provided by the dispersion model gives 4960 kg of flammable mass in the entire vapor cloud. The fraction of the flammable mass inside PESs (congested areas) is, then, 3470/4960 = 70%.

TABLE 7.14
Estimation of Flammable Mass in Congested Areas

Location of Cloud	x, m	Time, s	Δtime, s	Width Fraction in Congestion	Δm_f, kg
Start in Gas Plant	0	0			
Leave Gas Plant	50.3	8.6	8.6	1.0	400
Start in Cat Cracker	85.7	37.2			
Leave Cat Cracker	150	95.5	58.3	1.0	2711
Start in Turbo-Alt	174	115			
Leave Turbo-Alt	215	147	32	0.24	359
				Total	**3470**

Applying Step 5 the average heat of combustion in the cloud is found, which for a single-component surrogate, in this case 1-butene, is 45.30 MJ/kg.

Step 6 is to find the reactivity of the fuel as defined by the laminar flame speed. From Appendix C, the laminar flame speed for 1-butene ranges from 42.7 to 51 cm/s. These values fall along the lower boundary for medium reactivity fuel (Section 4.8.5).

Step 7 is to classify the degree of confinement for flame expansion for each PES. Using Section 4.8.6, areas in Figure 7.6 are designated 2D.

Step 8 is to classify a level of congestion by the Baker–Strehlow method (low, medium or high). In this case refinery structures are taken as medium congestion.

Step 9 is to use the classifications for Steps 6 through 8 with the Baker–Strehlow Table 4.8 to find the overpressure parameter, which is found to be $M_w = 0.662$. Table 7.15 summarizes the estimated explosion conditions.

TABLE 7.15
Estimated Explosion Conditions for LaMede Refinery Incident

Flame expansion	2D
Fuel reactivity	Medium
Obstacle density	Medium
Peak overpressure parameter, M_w	0.662

Consequently, the Sachs scaling factor used in the Baker–Strehlow explosion model is

$$\frac{E}{P_0} = \frac{m_f H_{comb}}{P_0} = \frac{3470 \times 45.30E6}{1.01E5} = 1.572E6\ m^3$$

$$\left(\frac{E}{P_0}\right)^{1/3} = 116.2\ m$$

The explosion overpressure curve from Figure 4.55A is read at overpressure values of 10, 5, and 3 psi and descaled to give the overpressure radii in Table 7.16.

These radii are plotted as dashed lines in Figure 7.1 for comparison with the damage estimates by the Total investigation team plotted as the solid circles. The choice of these overpressures is necessarily an over simplification to specifying damage levels. The agreement is seen to be very good. It is, however, based on our assumption of a 10-min liquid release.

TABLE 7.16
Finding Overpressure Radii for LaMede Refinery Example Explosion

ΔP Psig	ΔP Bar	$\frac{R}{(E/P_0)^{1/3}}$	R, m	Interpretation
10	0.697	0.47	55	Total destruction
5	0.348	0.82	95	80% damage
3	0.209	1.7	198	Damage threshold

7.5. Example 14, Partially Confined Vapor Cloud—Multiple Congested Areas

An example in which the vapor cloud overlaps more than one congested area is shown in Figure 7.9. This example is obtained by taking the 4-in. VCM pipe break (Example 11) at a different location and using a different wind direction. Now the cloud footprint passes through two 3D, *H* areas, a small portion of two 2D, *M* areas, and a 3D, *L* area near the source of the cloud.

As discussed in Sections 4.8.1 and 5.3.2, adjoining congested areas are merged for the purpose of explosion prediction. The distance between the 2D,M area and the large 3D,H area overlapped by the vapor cloud in Figure 7.9 is 6 m, so the two are considered separate explosion sources.

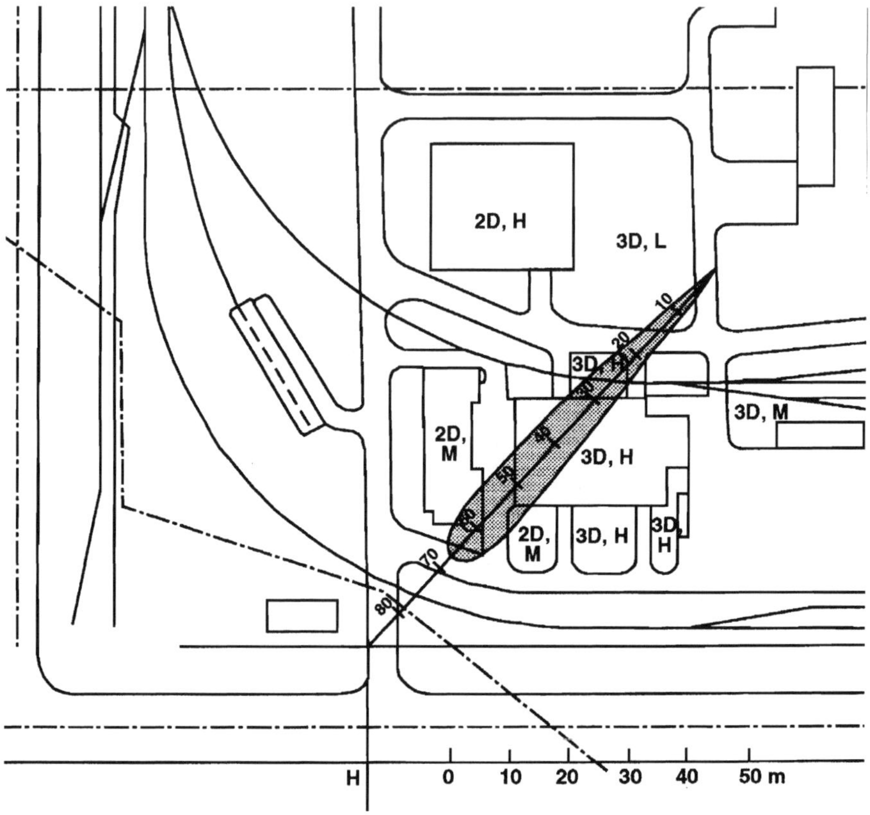

Applying Method 2 of Table 7.1 to allocate the distribution of flammable mass, the vapor cloud footprint areas in Figure 7.9 are defined for analysis purposes as follows.

1. Near source: from $x = 0$ to 22 m, in 3D, L area
2. Body: from $x = 22$ to 55 m, in small and large 3D, H areas and corner of 2D, M area which is lumped together with 3D, H area.
3. Nose: from $x = 55$ to 70.7 m

The nose area can be further distinguished by the portion overlapping the left-most 2D,M area (nose in) and the balance (nose out).

The vapor cloud is allocated to the above areas as shown in Table 7.17. In this case only a small part of the nose of the vapor cloud overlaps the 2D,M area. The flammable mass is 26.1 kg, or 4.5% of the total flammable mass in the cloud.

Summarizing, the explosion scaling parameters for this example as found from Tables 4.4 and 4.5 are listed in Table 7.18. From these parameters, the

TABLE 7.17
Allocation of Vapor Cloud Areas in Figure 7.9

Cloud Section	Centerline distance (m)	Area (m^2)	Average Height (m)	Volume (m^3)	Average Density (kg/m^3)	$\rho_{cld} - \rho_{air}$ (kg/m^3)	Flam. Mass (kg)
Near source	0–22	14.8	3.2	47.4	5.852	4.679	221.8
Body	22–55	93.4	5.6	523.0	1.64	0.467	244.2
Nose In	55–71	8.6	12.25	105.4	1.42	0.247	26.1
Totals		**145.3**		**1024.9**			**492.1**

TABLE 7.18
Summary of Explosion Parameters for Example Case

Cloud Section	Flam. Mass (kg)	Energy Scaling Factor, ES	Confinement, Congestion	Baker–Strehlow Parameter	Multi-Energy Parameter
Near source	221.8	34.35	3D, L	0.206	2-3
Body	244.2	35.46	3D, H	0.588	4-5
Nose In	26.1	16.83	2D, M	1.24	3-5

explosion overpressure and impulse curves can be readily obtained as illustrated in the previous example.

7.6. Example 15, Confined Vapor Clouds

The model of Cleaver et al. (1994) described in Section 5.4.2 is useful to discover whether the room or building under consideration will develop a stratified vapor cloud or a well-mixed cloud. It can be used in conjunction with a jet dispersion model as will be demonstrated. Consider an enclosure of width 30 m, length 40 m, and height 20 m. A 1-in. liquid horizontal propane line from a storage tank 5 m away is assumed to be sheared as it enters the building at an elevation of 2 m. If the storage tank is at 25 C, the vapor pressure of 9.5 bar(a) will cause a two-phase discharge rate of 2.20 kg/s of propane at an expansion velocity of 208 m/s with 66 mass % liquid droplets of Sauter mean diameter 7 μm. Dispersion modeling indicates that a jet of this aerosol will reach the conditions indicated in Table 7.19 as the centerline concentration dilutes to the LFL.

TABLE 7.19
Conditions of Aerosol Jet of Propane Diluted to the LFL

x distance (m)	Centerline height (m)	Plume width (m)	Plume depth (m)	Average density (kg/m^3)	Centerline velocity (m/s)
26.5	0.51	2.57	3.21	1.313	7.0

Taking the plume conditions at this point, an effective plume radius r_0 is 1.62 m since

$$\text{Cross-sectional area of plume} = 2.57 * 3.21 = \pi(1.62)^2$$

The momentum force for this plume is

$$M_c = \rho_{cld} u^2(\text{Width} * \text{Depth}) = (1.313)(2.57)(3.21)(7.0)^2 = 531 \text{ N}$$

The buoyancy force is

$$B_c = g\,\Delta\rho(Ar_0) = (30 * 40)\ 9.81(1.62)(1.313 - 1.173) = 2671 \text{ N}$$

The ratio of these forces is the overturning number:

$$\Omega = M_c/B_c = 531/2671 = 0.20$$

The effective gravitational constant is [Eq. (5.4.1)]

$$g_0' = \frac{g|\rho_a - \rho_g|}{\rho_g} = 9.81\ (1.313 - 1.173)/1.313 = 1.046 \text{ m/s}^2$$

The volume Richardson number is, by Eq. (5.4.4)

$$Ri_v = \frac{V^{1/3} g_0'}{u_0^2} = (24{,}000)^{1/2}\ 1.046/(7.0)^2 = 0.615$$

For a horizontal release, the additional depth of a well-mixed layer, δ by Eq. (5.4.5) is

$$\frac{\delta}{r_0} = \frac{C_i}{\sqrt{Ri_v}} = 40/0.785 = 50.96$$

or $\delta = 82$ m. For a vertical upward release, $C_I = 25$, and $\delta = 52$ m.

Since $\Omega > 0.1$, the formula for the mixing layer height, h, is, by Eq. (5.4.2):

$$h_1 = \min[0.5z_r + \delta, H] = \min[1 + 82, 20] = 20 \text{ m}$$

Thus, by a wide margin, the building in this example can be considered a well-mixed enclosure. Since the plume will be below the LFL past the point modeled so far, the flammable mass is that found by integration of the plume profiles, which in this case is 4.0 kg.

If the propane discharge goes on long enough, and assuming there is enough leakage from the building to avoid a build up of pressure, the well-mixed concentration will eventually increase above the LFL. Equation (5.4.8) applies with $c_0 = 1.0$ (pure propane entering the building.

$$c(t) = c_0\left[1 - e^{-\lambda t}\right] \qquad \text{for } t < t' \tag{5.4.8}$$

The volumetric rate of propane vapor after evaporating aerosol droplets would be

$$Q_0 = w/\rho_{\text{vap}} = 2.20/1.7848 = 1.233 \text{ m}^3\text{/s}$$

The time constant is $1/\lambda = V/Q_0 = 24{,}000/1.233 = 19{,}470$ s = 5.41 hr. Assuming t' is 10 min, the time it takes for $c(t)$ to reach the 2.1% LFL is found from Eq. (5.4.8) to be 6.89 min. After that time the flammable mass equals the inflow rate times time

$$m_\text{f} = 2.20t$$

until either ignition occurs or the concentration increases above the UFL.

8

RECOMMENDATIONS FOR FUTURE WORK

Below, distant, the roaring courtiers rise to their feet—
less shocked than irate.
Salome has dropped the seventh veil
and they've discovered there are eight.
—Danny Abse

In previous chapters, several issues arise which need further work to resolve. These issues are summarized here to stimulate research interests.

8.1. Calculating Flammable Mass Profiles Along a Vapor Cloud

Available computer models could improve their usefulness by reporting a flammable mass profile along the length of a vapor cloud. This would be particularly helpful when applying partially confined vapor cloud methods to predict explosion characteristics. In order to improve estimates of the flammable mass within each type of congested area, it would be useful to overlay portions of each congested area with a slice of the vapor cloud having a known flammable mass. These slices could then be readily added to obtain the flammable mass within each congested area.

8.2. Resolving the Minimum Explosive Mass Issue

Section 4.9 discusses the basis for expecting that vapor clouds containing less than some threshold minimum flammable mass will either not explode or will have a very low probability of ignition. This expectation can be verified or discredited with better experiments and models.

Explosion experiments have been conducted with small-scale obstacles to study the flame acceleration tendencies induced by plant obstacles. As the effect of obstacle scaling and obstacle density is better understood, we may be able to relate obstacle size and vapor cloud size to flame speed and explosion properties. A desirable outcome of such scaling studies is to find dimensions of full-scale equipment that represent a threshold for causing flame acceleration to velocities that cause damaging overpressures. An example of the apparent existence of such a threshold is given in Chapter 3 with the St. Herblain, France explosion. The conclusion of modeling studies analyzing this incident, is that five tanker trucks parked in parallel would not cause an explosion with gasoline vapors, but six will.

Risk analysts would like to have better justifications for setting a minimum threshold vapor cloud which would cause a damaging explosion. If there is a minimum flame run-up length to accelerate flames to explosive levels, and if this can be quantified as a function of type of plant equipment, then the accuracy and validity of explosion predictions will benefit.

8.3. Contribution of Aerosols to Explosive Mass

As discussed in Section 4.5, the LFL for aerosols can be less than half the LFL of a vapor of the same material (Cook et al., 1977). Furthermore, the minimum ignition energy has been shown to be lower for an aerosol than for a vapor. High flashpoint materials, when pumped, could leak under pressure and form a flammable aerosol (Bowen and Shirvill, 1994). Yet, there is a scarcity of measurement data on aerosol flammability limits and ignition energies. The few correlations that exist "all require exacting validation" (Bowen and Shirvill, 1994). It is also not clear what fraction of the liquid aerosol contributes to the energy released in an explosion.

8.4. Dispersion Modeling Around Plant Structure

It is acknowledged that models developed for open areas are not expected to accurately represent dispersion in highly congested plant areas. A few studies have been made using wind tunnels with scaled plant models (Peterson, 1987, 1990, 1993, 1997; Peterson and Ratcliff, 1988). A major study has been sponsored by the PERF group concerning the effect of obstacles on dispersion (Briggs et al., 1997). Generalizations are expected from such experimental programs. Much more is needed.

8.5. Improved Modeling of Jets Impacting Surfaces

The experimental work discussed in Section 4.2.5 needs to be incorporated into available dispersion models. Experimental verification of such models may require additional test data.

8.6. Models That Account for Turbulence Spectra

Current dispersion models focus almost entirely on predicting mean concentrations with various averaging time. Some computational fluid dynamic (CFD) models also calculate turbulence profiles and turbulence decay, most notably the K-ε models. Recently British Gas have produced a K-ε model of gas jets (Caulfield et al., 1993). Such models need to be made efficient and robust and available for widespread application.

8.7. Reconciling Indoor and Outdoor Explosion Models

Outdoor explosion models include the TNT equivalence method and the volume-source methods (Baker–Strehlow and multi-energy methods). Indoor explosion data have been used to specify explosion vent panel design in NFPA 68. The volume-source outdoor explosion models include parameters relating the degree of confinement. In a limit, indoor explosions represent a degree of confinement that should be treated by volume-source models. To fully reconcile the indoor and outdoor explosion test data explosion models like the volume-source models would have to be generalized to allow for the change of confinement that occurs when vent panels yield to open. Such a generalized model would very likely require an improved estimate of the flammable mass inside a building as discussed in Section 5.4.

8.8. Calculate Net Efficiencies for TNT Equivalent Models from Historical Events

Up-to-date dispersion models should be used to improve estimates of the flammable mass involved in explosion events. Then recalculate the explosion efficiency as a net efficiency. See if net explosion efficiencies are more nearly consistent than gross efficiencies previously used.

APPENDIX A

ATMOSPHERIC STABILITY CLASSIFICATION SCHEMES

Atmospheric stability is defined in terms of the tendency of a parcel of air to move upward or downward after it has been displaced vertically by a small amount. This is described in Hanna, Briggs, and Hosker (1982). Essentially, unstable atmospheres of Stability Class A tend to develop vertical updrafts which increase boundary-layer turbulence intensity. Stable atmospheres (Stability Class F) tend to suppress vertical updrafts and reduce turbulence intensity. Since it is difficult to measure turbulence intensity directly, correlations are sought to indicate stability class as a function of readily measurable variables.

The earliest stability classification scheme, attributed to Pasquill (1961), is summarized in Table A.1. This simply requires an estimate of solar radiation and wind speed. It has been shown to produce inconsistent classifications in comparison with other classification schemes during the Thorney Island dispersion tests (McQuaid, 1981).

TABLE A.1
Correlation of Atmospheric Stability Class According to Gifford (1976)

Wind Speed, m/s	Solar Insolation Strong	Solar Insolation Moderate	Solar Insolation Slight	Night Time Thin overcast or >1/2 low clouds	Night Time Conditions <3/8 cloudiness
<2	A	A-B	B	-	-
2-3	A-B	B	C	E	F
3-4	B	B-C	C	D	E
4-6	C	C-D	D	D	D
>6	C	D	D	D	D

If a meteorological tower is available it is possible to obtain measurements of both the vertical temperature profile and the vertical wind speed profile. These measurements are needed to evaluate the Richardson number, *Ri*, which is the ratio of the vertical temperature gradient to the squared vertical gradient of the wind speed. The numerator is related to the destabilizing forces that generate updrafts. The denominator is related to the kinetic energy that destroys updrafts. A larger value for the Richardson number should indicate more unstable atmospheres. Unfortunately, *Ri* varies with height, z. A correlation for stability class with Richardson number is given in Table A.2

Several Richardson numbers are in use. The one used in Table A.2 is designated by Sedefian and Bennett (1980) as the Businger version. It is defined below in finite difference form:

$$Ri = \frac{g}{T(z_1)} \frac{\left[\dfrac{T(z_1)-T(z_2)}{z_1 - z_2}\right]}{\left[\dfrac{u(z_1)-u(z_2)}{z_1 - z_2}\right]^2} \tag{A.1}$$

The potential temperature, θ, can be used instead of the actual temperature, T, to better account for the normal decrease in temperature with height, where

$$\theta = T - \lambda(z_2 - z_1) \tag{A.2}$$

and λ is the adiabatic lapse rate, ≈–0.01 K/m over the distances to the mixing height.

A simpler correlator that requires only vertical temperature information is the vertical temperature gradient, $\Delta T/\Delta z$. Theory relates the vertical temperature

TABLE A.2
Atmosphere Stability Class Correlated with Richardson Number

Pasquill Stability Class	$Ri = \dfrac{g(\Delta T/\Delta z)}{T(\Delta u/\Delta z)^2}$
A	$Ri < -0.86$
B	$-0.86 \le Ri < -0.37$
C	$-0.37 \le Ri < -0.10$
D	$-0.10 \le Ri < 0.053$
E	$0.053 \le Ri < 0.134$
F	$0.134 \le Ri$

TABLE A.3
Atmospheric Stability Class Correlated with Vertical Temperature Gradient

Pasquill Stability Class	$\Delta T/\Delta z$ (degrees K/100 m)
A	$\Delta T/\Delta z < -1.9$
B	$-1.9 \leq \Delta T/\Delta z < -1.7$
C	$-1.7 \leq \Delta T/\Delta z < -1.5$
D	$-1.5 \leq \Delta T/\Delta z < -0.5$
E	$-0.5 \leq \Delta T/\Delta z < 1.5$
F	$1.5 \leq \Delta T/\Delta z < 4.0$
G	$4.0 \leq \Delta T/\Delta z$

gradient to atmospheric stability when wind speeds are low. Table A.3 gives this correlation for each stability class.

An algorithm listed in Table A.4 has been adopted by the US Nuclear Regulatory Commission (NRC) as originated by Islitzer and Slade (1968). This is based on the standard deviations in the horizontal wind direction, σ_θ. Bivane anemometors record both σ_θ and the corresponding variations in the vertical component of wind speed, σ_ϕ . Since σ_θ varies with height; the values listed in Table A.4 apply to heights up to 10 m.

TABLE A.4
Atmospheric Stability Class Correlated with Standard Deviations of Horizontal Wind Direction

Pasquill Stability Class	σ_θ at 10 m, degrees
A	$\sigma_\theta > 22.5$
B	$22.5 \geq \sigma_\theta > 17.5$
C	$17.5 \geq \sigma_\theta > 12.5$
D	$12.5 \geq \sigma_\theta > 7.5$
E	$7.5 \geq \sigma_\theta > 3.75$
F	$3.75 \geq \sigma_\theta > 2.0$
G	$2.0 \geq \sigma_\theta$

Gifford (1976) reviewed stability typing schemes and preferred a scheme which uses the Monin–Obukov length, L, defined by

$$L = -\frac{\rho_{air} C_p T u_*^3}{k g H_v} \tag{A.3}$$

Here u_* is the friction velocity, C_p is the heat capacity of air, ρ_{air} is the ambient air density, T is the ambient temperature at the reference height, k is von Karman's constant (0.4), g is the gravitational constant, and H_v is the vertical heat flux. The difficult measurement to obtain is, of course, H_v. The correlation developed by Pasquill and Smith (1971) for flow over short grass, with z_o = 0.01 m is given in Table A.5.

TABLE A.5
Pasquill–Gifford Stability Class Correlated with Monin–Obukov Length

Stability Class	L, m
A	–2 to –3
B	–4 to –5
C	–12 to –15
D	∞
E	35 to 75
F	8 to 35

APPENDIX B

VERTICAL WIND PROFILES

Two alternative forms expressing wind profile as a function of height are

Power Law Form

$$u_{\text{w}}(z) = u_{\text{w}}(10)\left(\frac{z}{10}\right)^{p} \tag{B.1}$$

Logarithmic Form

$$u_{\text{w}} = \frac{u_*}{k}\left[\ln\left(\frac{z}{z_0}\right) - \Psi\left(\frac{z}{L}\right)\right] \tag{B.2}$$

where

u_* = Friction velocity, m/s
k = 0.40 = Von Karman constant
z_0 = Surface roughness length, m
L = Monin–Obukov turbulence scale, m

The Businger (1971) relationship is used for $\Psi(z/L)$ in terms of a characteristic mixing length L, the Monin–Obukov length:

$$L = \frac{\rho_{\text{air}} C_{\text{p}} T u_*^3}{k g H_{\text{v}}} \tag{B.3}$$

Spicer and Havens (1990) fit data by Pasquill (1974) to relate the Monin–Obukov length, L, to surface roughness length, z_0, giving

STABILITY	FUNCTION
A	$L = -11.4 z_0^{0.10}$
B	$L = -26.0 z_0^{0.17}$
C	$L = -123 z_0^{0.30}$
D	$L = 0$
E	$L = 123 z_0^{0.30}$
F	$L = 26.0 z_0^{0.17}$

For Stability A, B, C

$$\Psi = 2\ln\left(\frac{1+a}{2}\right) + \ln\left(\frac{1+a^2}{2}\right) - 2\arctan a + \frac{\pi}{2} \tag{B.4}$$

or, in general

$$\Psi = 2\left[\frac{(1+a)^2(1+a^2)}{(1+a_0)^2(1+a_0^2)}\right] - 2[\arctan a - \arctan a_0] \tag{B.5}$$

where

a_o = 1 so arctan $a_0 = \pi/4$
$a = (1 - 15z/L)^{0.25}$
and angles are expressed in radians.

Stability D

$$\Psi = 0$$

Stability E, F, G

$$\Psi = \beta z/L \tag{B.6}$$

where $\beta = 2.0$ is recommended by Irwin and Cope (1979), as a better fit to experimental data than the usually cited value, $\beta = 4.7$.

A value of the coefficient p in the power law relationship can be found that will provide a good fit between the power law relationship and the logarithmic relationship as described by Irwin and Cope (1979).

APPENDIX C

FLAMMABILITY PROPERTIES

Material	Formula	Mole Wt. (kg/kmole)	Boiling Pt. (°K)	Flammable Limit (% v/v)		Stoich in air (% v/v)	Auto Ignition Temp. (°K)	Flash point Closed Cup (°K)	Min. Ignition Energy in air (mJ)	Fundamental Burning Velocity S_u (cm/s)	Heat of Vaporization (kJ/kg) $k = 10^3$	Net Heat of Combustion (MJ/kg) $M = 10^6$
				Lower	Upper							
Default Reference		(d)	(d)	(d)	(d)		(d)	(j)				(d)
HIGH REACTIVITY												
Acetylene	C_2H_2	26.04	189d,n	2.4n 2.5d,h	80.0 100h	7.72n,p	578d,h	255d	0.017k,t	157q 166c,f 173e	250.8	48.22
Chlorine monoxide	Cl_2O	67.45	Explodes @ 282.9k	23.5j	100j	—						
Ethylacetylene	C_4H_6	54.09	281.2 350j	2.0d,j	11.5j 32.9		699j	269				40.02
Ethylene oxide	C_2H_4O	44.05	283.8d,n	3a,d	100a	7.72p	702d,r –713r	253d,r	0.062k,m 0.06t	89.5q 108c,f	581.1n 648.1	27.65

Material	Formula	Mole Wt. (kg/kmole)	Boiling Pt. (°K)	Flammable Limit (% v/v)		Stoich in air (% v/v)	Auto Ignition Temp. (°K)	Flash point Closed Cup (°K)	Min. Ignition Energy in air (mJ)	Fundamental Burning Velocity S_u (cm/s)	Heat of Vaporization (kJ/kg) $k = 10^3$	Net Heat of Combustion (MJ/kg) $M = 10^6$
				Lower	Upper							
Hydrogen	H_2	2.016	20.39 20.45n	4.0d,r 4.1r	72.5a 74.2r 75.0d,r	29.5p	673 773r 833r 844.3n		0.011r 0.017k, m,t 0.019c	265.18p 312c,f 320e	451.0n	119.95 119.9n
1,2 Propylene oxide	C_3H_{6O}	58.08	307.1	2.1 2.3r 2.8r	21.5	4.97p	722j,r 737r 738.2	235.9 d,r	0.13r 0.14k,m,t	67.2q 82c,f	525.6	30.74
1,3 Propylene oxide	C_3H_6O	58.08	321	2.8	37	4.97p	722j	245d	0.14k,m,t	67.2q 82f	525.6	31.01
AVERAGE REACTIVITY												
Acetaldehyde	C_2H_4O	44.05	216.5n 294d,h	1.6 2.6a	10.4 36a 60h	7.73	448h 458	235d,h	0.38k,m,t	140.21p	567.4n 585.7i	25.07 25.11a
Acetone	C_3H_6O	58.08	329.4 329.9n	2.5h 2.6d,m 2.9n	11.6n 12.5h 12.8 13a	4.97	810.9 834.3n	255d	1.15k,m,t	42.3q 45.72p 54c,f	523i,n	28.57 29.07a 30.8n
Benzene	C_6H_6	78.11	353d,h,n	1.2n 1.3a,h 1.4d,m	7.1 7.9a 8.0m 9.3n	2.72	771r 833r 835a,d 864.9n	262d,h	0.20r 0.22c,k, m,t 0.23r	39.62p 40.7e,s 47.8q 48c,f	431.8n 549.3	39.9n 40.14 40.63a

Material	Formula	Mole Wt. (kg/kmole)	Boiling Pt. (°K)	Flammable Limit (% v/v)		Stoich in air (% v/v)	Auto Ignition Temp. (°K)	Flash point Closed Cup (°K)	Min. Ignition Energy in air (mJ)	Funda-mental Burning Velocity S_u (cm/s)	Heat of Vapor-ization (kJ/kg) $k = 10^3$	Net Heat of Com-bustion (MJ/kg) $M = 10^6$
				Lower	Upper							
1-3 Butadiene	C_4H_6	54.09	269d,h	2.0d,h,r 2.1r	11.5d, h,r 12.5r	3.67	693h,r 702 703r	203r	0.13m,r,t 0.19r	52f 54.5q 54.86p 64c	440.0	44.55 44.64a
C4 = *n*-Butane	C_4H_{10}	58.12	272.7d,n	1.5r 1.68n 1.8	8.4r 8.5 10.3n	3.12	560r 638r 645r 678 703.8n	213	0.25c,k, m,t	36.58p 37.9q,s 40.5e 45c,f 45.9q	385.8n 484.3	45.72
1-Butene	C_4H_8	56.11	266.9j	1.6d,j,r 2.0n	9.3d,r 10j,r 13.3n	3.37	657 658j,r 716.5n 728r	193r	0.368r	42.67p 43.2q,s 50f 51c	443.9n 447.4	45.3d,n 45.39a
2-Butene-*cis*	C_4H_8	56.11	277	1.6 1.7j,r 1.8r	9.0j,r 9.7 9.8r	3.37	598d,j	261r			477.4	45.17
2-Butene-*trans*	C_4H_8	56.11	274	1.8d,j	4.7j 9.7d,r	3.37	597d,j	<267r			464.0	45.07
Carbon disulphide	CS_2	76.14	319.3d,n	1.2n 1.25r 1.3d,h	44r 50d,h 60r 73.0n	6.53	363d,h,r 375r 393.2n	241r 243d, h,r	0.009r 0.015k, m,t	48.77p 58c,f	351.0n 373.1	14.50

Material	Formula	Mole Wt. (kg/kmole)	Boiling Pt. (°K)	Flammable Limit (% v/v) Lower	Flammable Limit (% v/v) Upper	Stoich in air (% v/v)	Auto Ignition Temp. (°K)	Flash point Closed Cup (°K)	Min. Ignition Energy in air (mJ)	Fundamental Burning Velocity S_u (cm/s)	Heat of Vaporization (kJ/kg) $k = 10^3$	Net Heat of Combustion (MJ/kg) $M = 10^6$
Carbon oxysulfide	COS	60.08	223	12d,j	29d,j	10.69	—			46f	347.6	9.126
1-Chloro-propylene	C_3H_5Cl	76.53	308-309j	4.5j	16j	4.98		<267				
2-Chloro-propylene	C_3H_5Cl	76.53	296j	4.5j	16j	4.98	—	<253				
Cyanogen	C_2N_2	52.04	252d,h	6.0 6.6h	32d,h	9.48	—				530.8	21.06
Cyclohexane	C_6H_{12}	84.16	353.8d, n 355h	1.1n 1.2r 1.3d,h	7.8r 8.0d,b 8.4r 9.1n	2.27	518h,r 528–532b 533 543.2n	253d 256r	0.22m,r,t 0.24r	38.7q,s 43.6q 46c,f	258.1n 389.2	43.44 43.8n
Cyclopropane	C_3H_6	42.08	238.8n 240.4	2.4d,j 2.5n	10.4d,j 12.3n	4.45	771d,j,n		0.17r 0.18k,t	47.5q 48.77p 55.3q 56c,f	586.4	46.56
n-Decane	$C_{10}H_{22}$	142.3	447.2n 447.3	0.6n 0.8d,h	4.7n 5.4d,h	1.33	474r 478r 480.9 483h 504.9n	319b, d,h		43c,f	320.8 359.8n	44.2n 44.23

Material	Formula	Mole Wt. (kg/kmole)	Boiling Pt. (°K)	Flammable Limit (% v/v)		Stoich in air (% v/v)	Auto Ignition Temp. (°K)	Flash point Closed Cup (°K)	Min. Ignition Energy in air (mJ)	Fundamental Burning Velocity S_u (cm/s)	Heat of Vaporization (kJ/kg) $k = 10^3$	Net Heat of Combustion (MJ/kg) $M = 10^6$
				Lower	Upper							
Dichlorosilane	$H_2Si_2Cl_2$	101.0	281.5	4.0 4.1j,r 4.7h	96d,h 99j,r	9.48	309j,r 317h	221h 238j,r				81.93
Difluoroethane	$C_2H_4F_2$	66.05	247.4	3.7	18.0	7.73	—					11.48
Dimethylamine	C_2H_7N	45.08	280	2.8d,h,j	14.4d,j	5.29	673d,j,r 703h	223.2d 267r			617.8	35.82
Dimethyl ether	C_2H_6O	46.07	248.3 249j	3.4a,d,j 3.7r	18 27a,j	6.53	623d,j		0.29m,r,t 0.34r	54c	491.3	28.7a 28.83
2,2-Dimethyl-propane	C_5H_{12}	72.15	282	1.4d,j	7.5d,j	2.55	723d,j	<266r	1.570r,t	33.3q,s 33.53p 39c,f	327.6	45.05
Ethane	C_2H_6	30.07	184.3n 184.4h 184.6	2.8n 3.0	12.5d,r 15.3n 15.5r	5.65	745j,n,r 783r 788d,r	138	0.24c 0.25k,m,r,t 0.29r	39.62p 40.1e,s 47c,f	488.3n 521.5	47.4n 47.51
Ethylamine	C_2H_7N	45.08	290d,n	3.5d,h	13.5r 14d,h,r	5.29	653r 657 658h,j,r	<255r 277d	2.4m,r,t		611.3n 635.0	35.22
Ethylbenzene	C_8H_{10}	106.17	409	0.8j 1.0a,d,h	6.7a,d,h,j	1.96	705d,h,j	288.2d			366.6	40.92 41.38a

Material	Formula	Mole Wt. (kg/kmole)	Boiling Pt. (°K)	Flammable Limit (% v/v) Lower	Flammable Limit (% v/v) Upper	Stoich in air (% v/v)	Auto Ignition Temp. (°K)	Flash point Closed Cup (°K)	Min. Ignition Energy in air (mJ)	Fundamental Burning Velocity S_u (cm/s)	Heat of Vaporization (kJ/kg) $k = 10^3$	Net Heat of Combustion (MJ/kg) $M = 10^6$
Ethyl chloride	C_2H_5Cl	64.51	285	3.6r 3.8d,j,r	15.4d,j	6.53	768r 792d,j,r	223d,j	2.37m		409.3	19.92
Ethylene	C_2H_4	28.05	169.3h 169.5	2.7d,h	36d,h	6.53	723 763h		0.07k,m,t 0.12c	68.8e, 68.3s 74.5q 80c	515.0	47.15
Ethylene dichloride	$C_2H_4Cl_2$	98.96	357	6.2a,d,h	15.9a,h 16.0	7.73	686a,d	286d,h 291	2.37m,t		336.0	11.17 47.72a
Ethyl ether	$C_4H_{10}O$	74.12	308	1.9a,d,j	36a,j 48.0	3.37	453d,j	228d,j	0.19m 0.20k,t		353.3i	33.77
Ethyl mercaptan	C_2H_6S	62.14	308	2.8d,j	18.0d,j,r 18.2r	4.45	568r 572 573j,r	224.8d 255j,r			453.1	27.93 34.20a
Ethyl nitrate	$C_2H_5O_3N$	91.07	361j	4.0j		6.53		283j,r				
n-Hexane	C_6H_{14}	86.18	331.2n 342	1.1j,n 1.2	7.4r 7.5d,j,r 8.64n	2.16	496a,r 498j 506r 507 533.8n	241 250r 251.5d,r	0.25c,r 0.288k,m 0.29t	38.5e,s 39.62p 46c 50f	325.0 364.9n	44.70
Isobutane	C_4H_{10}	58.12	261	1.8d,j	8.4d,j	3.12	733d,j	192r	0.25r 0.38r	36.58p		45.58

Material	Formula	Mole Wt. (kg/kmole)	Boiling Pt. (°K)	Flammable Limit (% v/v)		Stoich in air (% v/v)	Auto Ignition Temp. (°K)	Flash point Closed Cup (°K)	Min. Ignition Energy in air (mJ)	Fundamental Burning Velocity S_u (cm/s)	Heat of Vaporization (kJ/kg) $k = 10^3$	Net Heat of Combustion (MJ/kg) $M = 10^6$
				Lower	Upper							
2-Isobutylene	C_4H_8	56.11	266	1.8d,j	8.8 9.6j	3.37	738d,j			53f	333.2	44.98
Isopentane	C_5H_{12}	72.15	301	1.4d,j	7.6d,j	2.55	693d,j	<222j,r	0.21m,t	36.58p	375.2	44.91
Isoprene	C_5H_8	68.12	307	1.5j 2d,h	8.9j 9d,h	2.91	493d,h 668j	219d, h,j		45.72p	450.5	43.81
Isopropylamine	C_3H_9N	59.11	305	2.0 2.3r	10.4	3.84	675d,j	223r 236d,j 247r	2.0m,t	27.43p 30f 31c		36.48
Isopropyl chloride	C_3H_7Cl	78.54	308j 308.9	2.8d,j	10.7d,j	4.45	866d,j	241d,j	1.55m,t		365.1	23.62
Methylamine	CH_5N	31.06	266.8	4.9d,h, j	20.7d, h,j	8.52	703.2d,h	255r 273.2d			870.4	31.39
3-Methyl-1-butene	C_5H_{10}	70.13	293	1.5a,d,j	9.1a,d,j	2.72	638d,j	211d <266j,r		41.5q 49c,f		44.55 44.95a
2-Methyl-1-butene	C_5H_{10}	70.13	304	1.4	9.6	2.40	638	253r <266 266.5d		39.0q 46c,f		44.42
Methyl formate	$C_2H_4O_2$	60.05	305d,n	4.5j 5.0a,h 5.9	20 23a,h,j	9.48	722j,r 729d,h,r 743r	254d, h,j	0.40r		472.0n 489.7	15.28a 15.34

Material	Formula	Mole Wt. (kg/kmole)	Boiling Pt. (°K)	Flammable Limit (% v/v)		Stoich in air (% v/v)	Auto Ignition Temp. (°K)	Flash point Closed Cup (°K)	Min. Ignition Energy in air (mJ)	Fundamental Burning Velocity S_u (cm/s)	Heat of Vaporization (kJ/kg) $k = 10^3$	Net Heat of Combustion (MJ/kg) $M = 10^6$
				Lower	Upper							
1,3-Pentadiene (*cis*)	C_5H_8	68.12	317.2	1.6 2.0r	8.3r 13.1	2.91	—	230r 232d 244j,r		45.72p 46.5q 55c,f	449.3	43.88
1,3-Pentadiene (*trans*)	C_5H_8	68.12	315.2	1.6 2.0r	8.3r 13.1	2.91	—	230d 244		45.72p 46.5q 54c	449.3	43.79
n-Pentane	C_5H_{12}	72.15	309.2d,n	1.4d,n 1.5j	7.8d,j 8.0r 9.2n	2.55	516r 531a 533d,j 557.6n 558r	224b,r 225r 233d, j,r	0.22m,r,t	38.5q,s 39.62p 44.4q 46c,f	364.4n 382.5	44.98 45.3n
1-Pentene	C_5H_{10}	70.13	303.1 303.2n	1.3n 1.5d,j	8.7d,j 10.0n	2.72p	545a 545.9 548j 571.5n	255d, k,j		42.67p 50c,f	413.5	44.63 45.0n
2-Pentene, (E) (*cis*)	C_5H_{10}	70.13	310	1.4	10.6	2.72p	561	227.6d 253	0.18m,t	42.67p,s 51c		44.52
2-Pentene, (Z) (*trans*)	C_5H_{10}	70.13	309.5	1.4	10.6	2.72p	558	225d 253	0.18m			44.46
Propadiene	C_3H_4	40.07	238.7	2.1 2.6r	—	4.98	531			73.15p 87c		46.33

Material	Formula	Mole Wt. (kg/kmole)	Boiling Pt. (°K)	Flammable Limit (% v/v)		Stoich in air (% v/v)	Auto Ignition Temp. (°K)	Flash point Closed Cup (°K)	Min. Ignition Energy in air (mJ)	Fundamental Burning Velocity S_u (cm/s)	Heat of Vaporization (kJ/kg) $k = 10^3$	Net Heat of Combustion (MJ/kg) $M = 10^6$
				Lower	Upper							
Propane	C_3H_8	44.09	231n 231.1	2.0r 2.1d,j,n 2.2r	9.5d,j,r 11.4n	4.02	723d,j 739a 743r 760r 777.6n	<169	0.25c,m,t 0.48r	39.0q,s 39.62p 45e 46c,f	425.5n	46.3n 46.34
Propylene	C_3H_6	42.08	226j 255.4	2.0d,j 2.4a	10.3r 11a,d 11.1j 11.7r	4.45	728d,j 733	165.4d	0.24r 0.28k,m,r	52c 43.8s	467.1	45.76 45.85a
Propyne	C_3H_4	40.07	250	1.7d,j	39.9	4.98	—	Gas	0.11r	69.9q,s 70.10p 71.3q 82c,f	588.3	46.16
Silane	SiH_4	32.12	104j 161d,h	1.4h	96h	9.48	Pyrophoric	305 311				44.30
Styrene	C_8H_8	104.2	418.3 419j	0.9j 1.1a,d,h	6.1a,d 7.0h 8.0r	2.05	763a,d,h,j	303r 305d,r 309r 311			387.1	40.49

Material	Formula	Mole Wt. (kg/kmole)	Boiling Pt. (°K)	Flammable Limit (% v/v)		Stoich in air (% v/v)	Auto Ignition Temp. (°K)	Flash point Closed Cup (°K)	Min. Ignition Energy in air (mJ)	Fundamental Burning Velocity S_u (cm/s)	Heat of Vaporization (kJ/kg) $k = 10^3$	Net Heat of Combustion (MJ/kg) $M = 10^6$
				Lower	Upper							
Tetrafluoro-ethylene	C_2F_4	100.0	195h 195.5d 197j	10j 11d,h	50j 60d,h	17.3	453r 461h 473d,j,r	<173h				1.28
Tetramethyl-silane	$C_3H_{10}Si$					2.55		<273r				
Toluene	C_7H_8	92.14	383.8d,n	0.98n 1.2a,d,h 1.3m 1.4r	6.7r 7.0m 7.1a,d,h 7.3n	2.27p	753j 808r 809a,d 841n	277.6d 279r 280	0.26r 1.30r 2.5k,m,t	36.58p 41c	362.6i, n	40.52 40.9n 41.00a
Trichlorosilane	$HC_{13}Si$	135.5	305d,j	1.2	90.5	29.52	377.6d,r 503r	223r 245.4d 248r 259				3.716
Trifluorochloro ethylene	C_2F_3Cl	104.46	191.7d 245j	8.4j	16.0j	17.3						3.010
Trimethylamine	C_3H_9N	59.11	276	2.0d,j	11.6d,j 12r	3.84	463d,j	<252r 266.5d	0.75r		450.3	37.98
Vinylacetylene	C_4H_4	52.08	278	2.0j,r 2.2	100j	4.02	—	—	0.080r 0.082m,t	76.20p 89c	536.2	33.67
Vinyl chloride	C_2H_3Cl	62.50	259.8	3.6d,h,j 4a	22a 33d,h,j	7.73	745d,j	195d,h			419.3	18.53

Material	Formula	Mole Wt. (kg/kmole)	Boiling Pt. (°K)	Flammable Limit (% v/v)		Stoich in air (% v/v)	Auto Ignition Temp. (°K)	Flash point Closed Cup (°K)	Min. Ignition Energy in air (mJ)	Funda-mental Burning Velocity S_u (cm/s)	Heat of Vapor-ization (kJ/kg) $k = 10^3$	Net Heat of Com-bustion (MJ/kg) $M = 10^6$
				Lower	Upper							
Vinyl ethyl ether	C_4H_8O	72.11	309	1.7d,j	28d,j	3.67	473r 475d,j	227 227.6d				32.91
Vinyl fluoride	C_2H_3F	46.04	201	2.6d,j	21.7d,j	7.73	658				381.5	2.194
Vinylidene chloride	$C_2H_2C_{12}$	96.94	305	6.5j 7.3d,h	15.5j 16.0h	9.48	843d,h,j	245 255d				10.35
Vinylidene fluoride	$C_2H_2F_2$	64.04	187.5	5.5d,j	21.3d,j	17.3	913					10.805
Vinyl methyl ether	C_3H_6O	58.08	279	2.6	39.0	4.98	560d,j	217.2d				30.55
LOW REACTIVITY												
Ammonia	NH_3	17.03	239.7 239.9n	15a 16.0 17r	25.0 28a	21.8	810a 903r 924.3d, n,r		680r		1373.6 n	18.60
Bromotrifluor-ethylene	C_2BrF_3	160.9	270.7		37d	17.3d	—	2.16d				1.967
Carbon monoxide	CO	28.01	81.7 83.2n	10.0n 12h 12.5d,j	74d,j 75h 199.4n	29.5	843r 878r 882d,j,n 973h			39.62p 46c	211.7n 241.1	10.11

Material	Formula	Mole Wt. (kg/kmole)	Boiling Pt. (°K)	Flammable Limit (% v/v)		Stoich in air (% v/v)	Auto Ignition Temp. (°K)	Flash point Closed Cup (°K)	Min. Ignition Energy in air (mJ)	Fundamental Burning Velocity S_u (cm/s)	Heat of Vaporization (kJ/kg) $k = 10^3$	Net Heat of Combustion (MJ/kg) $M = 10^6$
				Lower	Upper							
Ethanol	C_2H_6O	46.07	351 351.7n	3.3a,j 4.3	19a,d,j	6.53	636j 665.4n 695.9 698r	285r 286d,j	0.40r		836.8n 878.6	26.82
Methane	CH_4	16.04	111.5n 111.7	4.4n 5.0d,j 5.3r	14r 15.0d,j 15.5a	9.48	810r 813m 868r 873 905.4n	Gas	0.28r 0.29c 0.30m,r,t	32.53p 33.8q,s 40c,f	509.2n 555.3	27.83a 50.02d,n
Methanol	CH_4O	32.04	337.7n 337.9	5.9n 6r 6.7a 7.3	36a,d 49.9n	12.25	658r 728r 737 743.2n	284.3d, r 285r	0.140r,t	57.2q	1100.9 n	19.8n 19.92
Methyl chloride	CH_3Cl	50.49	248.9	6.0j 7.6r 10.7	13.4r 17.4d,r 19r 36j	12.25	737j 898r 905d,r	223r 284			445.4	13.38 20.79a

References

a Zabetakis (1965)
b Coward and Jones (1952)
c Factory Mutual Engineering Corp. (1967)
d Daubert and Danner (1989)
e Fiock (1955)
f NFPA 68
g MSDS, 1990
h NFPA 49
i NFPA 30
j NFPA 325M
k CRC *Handbook of Chemistry and Physics,* 70th edition, 1984
m NFPA *Fire Protection Handbook,* 1991
n SFPE *Handbook of Fire Protection Engineering* (2nd ed), SFPE/NFPA, 1995
p Lees (1980)
q Hercules Data Guide, HE-118A, Hercules, Inc., St. Rte 956, Rocket Center WV.
r Bond, 1991
s Gerstein et al., 1950
t NFPA, *Fire Protection Handbook.* Table 11-3B, 15th ed. 1981.

APPENDIX D

CORRELATION FOR FLASH POINT

A correlation for flash point given by Satyanarayana and Rao (1992) uses the three-parameter model for flash point, T_f, in terms of temperature, T, and normal boiling point, T_b:

$$T_f = \frac{a + b(c/T_b)\exp(-c/T)}{[1 - \exp(-c/T)]^2} \tag{D.1}$$

The constants in Eq. (D.1) are listed in Table D.1 by chemical type, along with the average absolute error (AAE) for the materials tested.

TABLE D.1
Constants for Flash Point Correlation

Group	No. of compounds	a	b	c	AAE%
Hydrocarbons	230	225.10	537.60	2217.00	0.410
Alcohols	150	230.80	390.50	1780.00	0.186
Amines	70	222.40	416.60	1900.00	0.200
Acids	40	323.20	600.10	2970.00	0.531
Ethers	80	275.90	700.00	2879.00	0.343
Sulfur	40	238.00	577.90	2297.00	0.484
Esters	120	260.80	449.20	2217.00	0.186
Ketones	80	260.50	296.00	1908.00	0.722
Halogen compounds	200	262.10	414.00	2154.00	0.672
Aldehydes	45	264.50	293.00	1970.00	0.771
Phosphorus Compounds	20	201.70	416.10	1666.00	0.815
Non-amine Nitrogen Compounds	125	185.70	432.11	1645.00	0.622
Petroleum fractions	21	237.90	334.40	1807.00	0.294

APPENDIX E

POLYDISPERSE DROP SIZE DISTRIBUTIONS

Aerosol drop size distributions are usually found to be log normal, although drops formed by condensation tend to be more nearly monodisperse (or at least have a narrow distribution). The log normal distribution function $f(t)$ is given in terms of the drop diameter, d_p the mean drop size, d_{pm}, by

$$f(t) = \frac{\exp(-0.5x^2)}{\sqrt{2\pi}\,\ln\sigma_G}, \qquad x = \frac{\ln t}{\ln\sigma_G}, \qquad t = \frac{d_p}{d_{pm}}$$

Here, σ_G is the standard deviation of the distribution, a smaller number when the distribution is narrow. Typically the log-normal standard deviation ranges $1.3 < \sigma_G < 1.9$. The cumulative distribution is simply the integral of $f(t)$, denoted $F(t)$. Figure E.1 is an example of a log-normal drop size distribution with $\sigma_G = 1.9$. Grid lines are shown for the mean drop size that occurs near the peak of the $f(t)$ distribution.

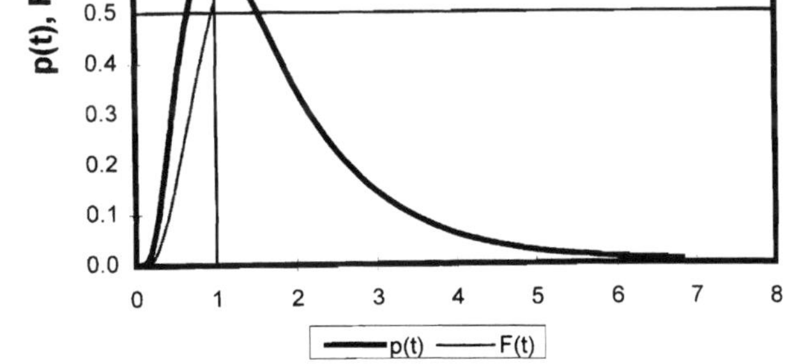

Figure E.1. Typical log normal distribution and cumulative distribution of drop sizes for σ_G of 1.9

APPENDIX F

MULTICOMPONENT POOL EVAPORATION FOR SPILLS ON LAND

For a multicomponent liquid spill or rainout, it is of interest to know both the total evaporation rate and the individual component evaporation rates. In turn, this gives the composition of the evaporating vapor. The methods developed for single component evaporation are readily extended to multicomponents. Spread rate formulas apply unchanged.

As developed by Wu and Schroy (1979) and Drivas (1982) by assuming ideal solution theory, Raoult's law, and a well-mixed pool a particularly simple solution results. Dividing the evaporation rate formula in its simplified form by the molecular weight, M_i, converts the component mass evaporation rate dm_i/dt to the component molar evaporation rate, dn_i/dt, and by the definition of mole fraction $y_{si} = n_i/n_T$, gives

$$\frac{dn_i}{dt} = -A_p \frac{k_g}{n_T} \frac{\rho_i(T_g)}{M_i} n_i = kn_i \tag{F.1}$$

Details are given in *RELEASE: A Model with Data to Predict Aerosol Rainout in Accidental Releases* (CCPS, 1998, Appendix C).

The ratio ρ_i/M_i is just the molar density, equal for all components. Since the total evaporation flux $dn_T/A_p dt$ is nearly constant, the n_T/A_p will be constant, and the coefficient of n_i will be a constant, k. The component evaportation rate integrates to

$$n_i = n_{i0} \exp(-kt) \tag{F.2}$$

The total evaporation rate is given by differentiating this expression for n_i and adding over all components

$$\frac{dm_T}{dt} = -k\sum_i^N M_i n_{i0} \exp(-kt) \tag{F.3}$$

Applying Raoult's law gives the total evaporation rate in terms of the initial mole fraction of each component in the liquid, x_{i0}, and the vapor pressure of each component at the initial liquid temperature, $P_{si}(T_{L0})$:

$$\frac{dm_{\mathrm{T}}}{dt} = -k\sum_{i}^{N} M_i x_{i0} P_{\mathrm{S}i}(T_{\mathrm{L0}}) \exp(-kt) \quad \text{(F.4)}$$

Improvements were made to this simplified model by Frie et al. (1992) by postulating a stratified pool with a thin surface layer that drops in temperature while the bulk of the liquid below remains at constant temperature.

In modeling evaporation from an aqueous solution of ammonia, Frie et al. found that liquid-phase resistance is important in this system for a short period (under 1 hr), and gas phase resistance dominates for the remainder of the evaporation. The two-layer model fits experimental data better, showing the importance of correctly modeling the initial liquid-phase controlling phase.

APPENDIX G

GENERALIZED INDOOR CONCENTRATION BUILD-UP OR DECAY

When a release of flammable materials occurs outdoors and can enter a building, it is possible to calculate the time-varying concentrations inside the building. The appropriate starting equation is the mass balance on a well-mixed building, (equivalent to the model for a continuous back-stirred reactor) following the notation in Figure 5.20.

$$V\frac{dc_i}{dt} = q_0 c_o (1 - F_0) + q_1 c_i (1 - F_1) + q_2 c_o - (q_0 + q_1 + q_2) c_i + s - R \qquad \text{(G.1)}$$

The first three terms on the right-hand side of Eq. (G.1) represent the rate at which the contaminant enters the building. The fourth represents the rate at which it leaves in output streams. For flammable components the indoor source , s, and sink, R, are zero since the component does not deposit onto the surfaces, etc, but they are retained here for generality. Likewise, the gaseous components of a vapor cloud are not usually captured by the intake filter, so the filter efficiency, F_0 and F_1 are zero. However, some aerosol droplets may be filtered out, so the F values are retained.

For simplification, Eq. (G.1) is made nondimensional by defining

$$\theta_i = c_i / c_{ref}, \quad \tau = q_0 t / V, \quad \alpha = q_1 / q_0, \quad \beta = q_2 / q_0,$$
$$\gamma = \sum_j k_j A_j / q_0, \quad \delta = 1 + \alpha F_1 + \beta + \gamma, \quad \varepsilon = 1 + \beta - F_0,$$
$$\theta_o = c_o / c_{ref}, \quad \text{and} \quad \sigma = s / q_0 c_{ref}$$

The mass balance reduces to

$$\frac{d\theta_i}{d\tau} + \delta\theta_i = \varepsilon\theta_o + \sigma \qquad \text{(G.2)}$$

The solution to this equation in general is

$$\theta_i = \varepsilon e^{-\delta\tau} \int_0^\tau e^{\delta\tau} \theta_o \, d\tau + e^{-\delta\tau} \int_0^\tau e^{\delta\tau} \sigma \, d\tau + \theta_{i0} e^{-\delta\tau} \tag{G.3}$$

where C_{i0} is the initial indoor concentration, usually zero, and $\theta_{i0} = c_{i0}/c_{ref}$. There are four time scales associated with the time constant for build up and decay of concentration, $1/\delta$ and τ. These are

- V/q_0 = time constant to exchange building air with makeup air
- $V/(q_1 F_1)$ = time constant to exchange building air with infiltrated air
- V/q_2 = time constant to remove the pollutant with the input filter
- $V/\sum_j k_j A_j$ = time constant to remove the pollutant via first-order decomposition mechanism

A practical analytic solution to Eq. (5.4.3) is to break up the outdoor concentration concentration vs. time profile into a series of ramp functions, that is straight line interpolations between data points. A single ramp function is given by

$$c_o = c_{o0} + at \tag{G.4}$$

The indoor concentration for outdoor concentrations varying along a single ramp function are

$$\theta_i \; \frac{\varepsilon\theta_{o0} + \sigma}{\delta} + \frac{b\varepsilon}{\delta^2}(\delta\tau - 1) + \left[\theta_{i0} + \frac{\varepsilon b}{\delta^2} - \left(\frac{\varepsilon\theta_{o0} + \sigma}{\delta} \right) \right] e^{-\delta\tau} \tag{G.5}$$

where $\theta_{o0} = c_o/c_{ref}$, $b = aV/c_{ref} q_0$, $a = (c_{o1} - c_{o0})/t_1$

The solution for m time intervals of length t_1 after the mth time interval is

$$\begin{aligned} c_{i(m)} &= S + \left[Ac_{o(m-1)} + Bc_{om} \right] \\ &+ \left[A(c_{o(m-2)} - c_{o(m-1)}) + B(c_{o(m-1)} - c_{o(m)}) \right] e^{-\delta\tau_1} + \\ &+ \left[A(c_{o(m-3)} - c_{o(m-2)}) + B(c_{o(m-2)} - c_{o(m-1)}) \right] e^{-2\delta\tau_1} + \cdots \\ &+ \left[A(c_{o(0)} - c_{o(1)}) + B(c_{o(1)} - c_{o(2)}) \right] e^{-(m-1)\delta\tau_1} \\ &+ \left[c_{i(0)} - S - Ac_{o(0)} - Bc_{o(1)} \right] e^{-m\delta\tau_1} \end{aligned} \tag{G.6}$$

where $c_{i(m)}$ is the indoor pollutant concentration after the mth time interval of length t_1, $c_{o(m)}$ is the outdoor pollutant concentration measured at the end of the mth time interval of length t_1, $S = s/q_0\delta$, $A = \varepsilon/\delta^2 t_1$, $B = (\varepsilon/\delta^2 t_1)(\delta\tau_1 - 1)$, $\tau_1 = q_0\tau_1/V$, and the rest of the terms are the same as those defined previously.

APPENDIX H

CALCULATING CONCENTRATIONS FOR INDOOR RELEASES

As explained in Section 5.4.2, a gas pipe break inside a relatively small room will produced a well-mixed concentration which gradually increases in time as approximated by the formulas in Appendix G. On the other hand, a gas injected into a large room with low ventilation will generate two layers of gas, one well-mixed, and the other with a vertical concentration profile that is linear in height. If the gas is lighter than air, the well-mixed layer will be the upper layer, and vice versa if the gas is heavier than air. The height of the well-mixed layer is h_1 and the height of the other layer is h_2 as shown in Figure 5.22. The depth of the well-mixed layer increases but the depth of the layer with the linear concentration profile remains about the same. Cleaver et al. (1994) proposed the model detailed below to describe this double-layer stratification.

The equations for conservation of mass in the two layers are written in terms of the average concentrations in each layer, c_1 and c_2, and the volumetric flow rate Q_0 and concentration c_0 from the pipe, and Q_1 across the interface between the two layers of cross-sectional area A. There is no flow Q_2 across the interface of the bottom layer, since that interface defines the layer and any flow would simply change the location of the interface. Since the volume of the upper layer and h_1 are constant, the volumetric flow into this layer, Q_1 equals the volumetric flow out of this layer. There is a volumetric flow rate $Q_3 = Q_0$ flowing out of the room (to avoid pressurization). Until the lower interface h_2 reaches the floor of the room the outflow Q_3 does not contain any flammable material, but afterwards it does.

The mass balance for layer 1 is

$$\frac{d(h_1 c_1)}{dt} = h_1 \frac{dc_1}{dt} = \frac{Q_1}{A}(c_2 - c_1) \tag{H.1}$$

The growth of layer 2 is governed by the volumetric balance on the layer, which since there is equal volumetric flow across the top of this layer, and no flow across the bottom is simply:

$$\frac{dh_2}{dt} = \frac{Q_0}{A} \tag{H.2}$$

This holds until h_2 reaches the floor of the room. After this time, h_2 is fixed at $h_2 = H_r - h_1$. The mass balance across layer 2 is

$$\frac{d(h_2 c_2)}{dt} = \frac{Q_0}{A} c_0 - \frac{Q_1}{A}(c_2 - c_1) = h_2 \frac{dc_2}{dt} + c_2 \frac{dh_2}{dt} \tag{H.3}$$

or, substituting in Eq. (H.2):

$$h_2 \frac{dc_2}{dt} = \frac{Q_0}{A}(c_0 - c_2) - \frac{Q_1}{A}(c_2 - c_1) \tag{H.4}$$

Equation (H.4) applies also after at $h_2 = H - h_1$, since there is an outflow of flammable material of rate $Q_3 c_2$ which is added to the mass balance. These equations predict the build-up of concentration over time. Ignition can occur at any point, so the flammable mass must be known at that time.

The flammable mass is found from that portion of the concentration that is above the LFL:

$$F_m = [c_1 V_1 + \text{portion of } c(z) V_2 \text{ above LFL}] \tag{H.5}$$

Modeling Details

To model this behavior, define the volume rate $Q_0 = \pi r_0^2 u_0$, where r_0 is the radius of the nozzle, the momentum flux is $M = Q_0 u_0$, the buoyancy flux is $B = g'_0 Q_0$, and the reduced gravity (buoyancy-adjusted gravity) g'_0 is

$$g'_0 = \frac{g|\rho_a - \rho_g|}{\rho_g} \tag{H.6}$$

The height of the well-mixed layer, h_1, depends on the ratio of stabilizing and destabilizing factors. Baines and Turner (1967) studied buoyant releases with negligible momentum inside a closed volume. They found that the behavior is governed by the ratio of the cloud momentum, $M_c = \rho \pi r^2 u^2$ and the cloud buoyancy flux, $B_c = Agr\Delta\rho$, where A is the horizontal cross-sectional area, and ρ, r, u, and $\Delta\rho$ denote the local density, radius, velocity, and density difference from ambient air of the plume when it reaches the ceiling. They defined the ratio of M_c

to B_c as the overturning number, Ω, and observed that overturning took place when

$$\Omega = M_c/B_c > 0.1 \qquad \text{(H.7)}$$

A stratified layer is produced which grows downward from the ceiling for smaller values of Ω.

Based on the Baines and Turner criteria, the depth of the mixing layer is modeled by

$$h_1 = \begin{cases} \min[0.5z_r + \delta, H] & \text{for } \Omega > 0.3 \\ \{(0.3-\Omega)\min[0.5z_r + \delta, H] + (\Omega - 0.1)0.5z_r / 0.2 & \text{for } 0.1 < \Omega < 0.3 \\ (\Omega/0.1)0.5z_r & \text{for } \Omega < 0.1 \end{cases} \qquad \text{(H.8)}$$

This relationship assumes that a purely buoyant source would produce a well-mixed layer extending to a fraction of the distance between the release and the ceiling, f_r. The additional depth, δ, of the well-mixed layer is attributed to the initial momentum of the release. Experimentally, the fraction of f_r that should be subtracted from h is ½, so δ is defined by

$$\delta = h - 0.5f_r \qquad \text{(H.9)}$$

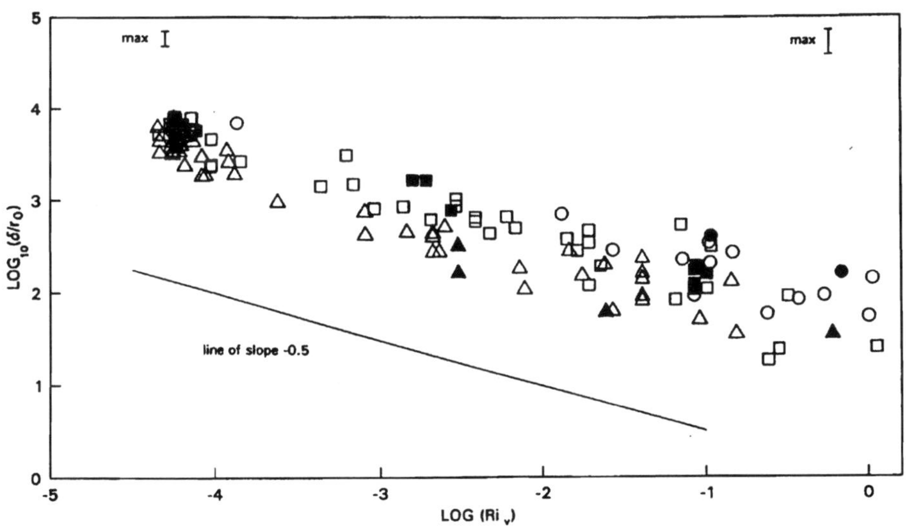

Figure H.1. Depth of transitional mixing layer as a function of release conditions (Cleaver et al., 1994) for methane injected into an enclosure. The open symbols represent full-scale experiments; the filled ones small-scale bring-water experiments. The release direction is vertical upward (O), horizontal (□), vertical downward (Δ). The error bars refer to the maximum uncertainty in the gas-air experiments. (Reproduced by permission of Elsevier Science Publishers B.V.)

Experimental data by Cleaver et al. are plotted in Figure H.1 as the ratio δ/r_0 against the inlet Richardson number, Ri_v defined by

$$Ri_v = \frac{V^{1/3} g_0'}{u_0^2} \tag{H.10}$$

The inlet Richardson number provides a measure of the ability of the jet to promote mixing in the volume since it represents the ratio of the reduced gravity (buoyancy-adjusted gravity) g_0' times the characteristic length of the room to the specific kinetic energy of the release (J/kg).

Cleaver et al. (1994) provide data indicating a relationship for the transition layer depth given by

$$\frac{\delta}{r_0} = \frac{C_i}{\sqrt{Ri_v}} \tag{H.11}$$

With C_i values of 25, 40, and 63 for vertically upward, horizontal, and vertically downward releases, respectively. The Richardson number represents the ratio of buoyancy to kinetic energy, so Eq. (H.11) represents a stability threshold for buoyant stratification or mixing.

APPENDIX I

EVALUATING FLAMMABLE MASS FOR GAUSSIAN DISPERSION MODELS: INSTANTANEOUS, POINT SOURCE

I.1. Integration Gives Flammable Mass

The general equation for finding the flammable mass between the lower and upper flammability limits is a triple integral over the volume of the cloud between the coordinates at the LFL, x_{LFL}, $\pm y_{LFL}$, $\pm z_{LFL}$:

$$m_F(t) = \int_0^{x_{LFL}} \int_{-y_{LFL}}^{y_{LFL}} \int_{-z_{LFL}}^{z_{LFL}} f(x, y, z, t)\, dx\, dy\, dz \quad \text{(kg)} \tag{I.1}$$

where the function $f(x, y, z, t)$ is defined as follows:

$$f(x, y, z, t) = \begin{cases} c(x, y, z, t) & \text{if LFL} < c(x, y, z, t) < \text{UFL} \\ 0 & \text{otherwise} \end{cases}$$

and the lower limit in z is 0 (ground) if $-z_{LFL}$ falls below ground level.

I.2 Gaussian Model Form

The concentration $c(x, y, z, t)$ in lb/ft^3 or kg/m^3 is expressed in terms of the discharge rate, w, the wind speed, u_w, and the lateral and vertical profiles by

Continuous (and Finite Duration) Source:

$$c(x, y, z) = \frac{w}{u_w} F_y(x, y) F_z(x, z) \quad (\text{kg m}^{-3}) \tag{I.2}$$

Instantaneous (Puff) Source

$$c(x, y, z, t) = w_p F_x(x, t) F_y(tu_a, y) F_z(tu_a, z) \quad (\text{kg m}^{-3}) \tag{I.3}$$

The cross-wind lateral dispersion for a *point source* is given in terms of σ_y which is a known function of x:

$$F_y(x, y) = \frac{1}{\sqrt{2\pi}\, \sigma_y(x)} \exp\left(\frac{y^2}{2\sigma_y^2(x)}\right) \quad (\text{m}^{-1}) \tag{I.4}$$

If the lateral width of the source is $2b_{0y} > 0$ use the *line source* expression:

$$F_y(x, y) = \frac{1}{4b_{0y}} \left\{ \text{erf}\left(\frac{b_{0y} - y}{\sqrt{2}\sigma_y(x)}\right) + \text{erf}\left(\frac{b_{0y} + y}{\sqrt{2}\sigma_y(x)}\right) \right\} \quad (\text{m}^{-1}) \tag{I.5}$$

The function $F_x(x, x_0)$ is the same as $F_y(x, y)$ with y replaced by $x - x_0$ where x_0 is the position of the center of the cloud for both point or line sources.

The error function erf(x) is tabulated in standard handbooks (CRC, 1997) and is defined as

$$\text{erf}(x) = \frac{2}{\sqrt{\pi} \int_0^x \exp(-t^2)\, dt}$$

A reasonable approximation given by Press et al. (1986) and the *Yellow Book* (1997) is given by

$$\text{erf}(x) = \frac{x}{|x|} \sqrt{1 - \exp\left(-\frac{4x^2}{\pi}\right)} \tag{I.6}$$

The vertical dispersion expression for a point source for values of

$$\sigma_z(x) \le 0.6 h_i \sqrt{1 - h/h_i}$$

is

$$F_z(x, z) = \frac{1}{\sqrt{2\pi}\, \sigma_z(x)} \left\{ \exp\left(\frac{(z-h)^2}{2\sigma_z^2(x)}\right) + \exp\left(\frac{(z+h)^2}{2\sigma_z^2(x)}\right) \right\} \quad (\text{m}^{-1}) \tag{I.7}$$

The latter two terms express the concept of reflected contours, or a folding back of the contour that would have gone below the ground level. When the source is $2b_{0y} > 0$ and $\sigma_z(x) \le 0.6 h_i \sqrt{1 - h/h_i}$, use a *line source* given by:

$$\begin{aligned} F_z(x, z) = \frac{1}{4b_{0z}} \Bigg\{ & \text{erf}\left(\frac{b_{0z} - z + h}{\sqrt{2}\sigma_z(x)}\right) + \text{erf}\left(\frac{b_{0z} + z - h}{\sqrt{2}\sigma_z(x)}\right) \\ & + \text{erf}\left(\frac{b_{0z} - z - h}{\sqrt{2}\sigma_z(x)}\right) + \text{erf}\left(\frac{b_{0z} + z + h}{\sqrt{2}\sigma_z(x)}\right) \Bigg\} \quad (\text{m}^{-1}) \end{aligned} \tag{I.8}$$

I.3. Solution for Instantaneous, Point Source

An analytic solution is available for the flammable mass in an instantaneous, point-source Gaussian plume. Integration of Eq. (I.3) using Eq. (I.4) for either a grounded plume or a totally elevated plume between the LFL dimensions $2x_{\mathrm{LFL}}$, $2y_{\mathrm{LFL}}$, and $2z_{\mathrm{LFL}}$ gives the ratio of flammable mass to total mass in the cloud, m, as (Van Buijtenen, 1980, citing Crank, 1956):

$$\frac{m_{\mathrm{f}}}{m} = \mathrm{erf}\left[\sqrt{\ln\left(\frac{c_0}{c_{\mathrm{LFL}}}\right)}\right] - \mathrm{erf}\left[\sqrt{\ln\left(\frac{c_0}{c_{\mathrm{UFL}}}\right)}\right] - \frac{2c_{\mathrm{LFL}}}{c_0\sqrt{\pi}}\sqrt{\ln\left(\frac{c_0}{c_{\mathrm{LFL}}}\right)} + \frac{2c_{\mathrm{UFL}}}{c_0\sqrt{\pi}}\sqrt{\ln\left(\frac{c_0}{c_{\mathrm{UFL}}}\right)} \quad \text{(I.9a)}$$

or if $c_o < c_{\mathrm{UFL}}$ it reduces to

$$\frac{m_{\mathrm{f}}}{m} = \exp\left[\sqrt{\ln\left(\frac{c_0}{c_{\mathrm{LFL}}}\right)}\right] - \frac{2c_{\mathrm{LFL}}}{c_0\sqrt{\pi}}\left[\sqrt{\ln\left(\frac{c_0}{c_{\mathrm{LFL}}}\right)}\right] \quad \text{(I.9b)}$$

For this idealized case, the maximum ratio of gas released that can be in the flammable region at one time is determined only by the ratio of the UFL to the LFL. This is shown by differentiating Eq. (I.9a) with respect to c_0, showing that the maximum value of m_{f}/m is reached when the following is satisfied:

$$\ln c_0 = \frac{c_{\mathrm{UFL}}^2 \ln c_{\mathrm{UFL}} - c_{\mathrm{LFL}}^2 \ln c_{\mathrm{LFL}}}{c_{\mathrm{UFL}}^2 - c_{\mathrm{LFL}}^2} \quad \text{(I.10)}$$

Before substituting this result back into Eq. (I.9), first use the transformations

$$\text{if } v = \frac{c_{\mathrm{UFL}}}{c_{\mathrm{UFL}}} \qquad v_1 = \frac{\ln v}{v^2 - 1} \qquad v_2 = \frac{v^2 \ln v}{v^2 - 1} \quad \text{(I.11)}$$

which leads to

$$\left(\frac{m_{\mathrm{f}}}{m}\right)_{\max} = \exp\left(\sqrt{v_2}\right) - \exp\left(\sqrt{v_1}\right) + \frac{2}{\sqrt{\pi}}\left[\exp(-v_1)\sqrt{v_1} - \exp(-v_2)\sqrt{v_2}\right] \quad \text{(I.12)}$$

This equation is plotted in Section 5.2.2, Figure 5.10. A solution is also given for a line source.

APPENDIX J

EVALUATING FLAMMABLE MASS FOR GAUSSIAN DISPERSION MODELS: CONTINUOUS RELEASE, APPROXIMATE METHOD

The *Yellow Book* (Committee for the Prevention of Disasters, 1997, pp. 4.68–4.76) method for estimating flammable mass is detailed below. The*Yellow Book* makes use of a modern form for the σ functions.This form provides flexibility in treating both urban and rural conditions via a dependence on the surface roughness length, z_0. First we need to define the terms used in their Gaussian model.

J.1. Standard Deviation Formulas

The *lateral* standard deviation $\sigma_Y(x)$ is given in terms of the wind speed, u_a, the cross-wind turbulent velocity, σ_v, and a time scale for turbulence, t_i as

$$\sigma_y = \sigma_v \frac{x}{u_a} \frac{1}{1+0.9\sqrt{x/u_a t_i}} \quad \text{(m)} \tag{J.1}$$

The time scale is given a value depending on the ratio of the source height, h, to the mixing height (height of an inversion layer trapping atmospheric pollutants) h_i:

$$t_i = \begin{cases} 300 \text{ s}, & h/h_i < 0.1 \\ 1000 \text{ s}, & h/h_i \geq 0.1 \end{cases} \tag{J.2}$$

The cross-wind turbulent velocity, σ_v, also depends on the mixing height, and also the friction velocity, u_*, which is a parameter in the vertical wind speed profile, the Monin–Obukov length, L, which depends on atmospheric stability class, and the Von Karman constant, k = 0.35 or 0.4.

$$\begin{cases} \sigma_v = 1.9u_* \sqrt{1-\dfrac{h}{h_i}} < 0.1 & \text{for } \dfrac{1}{L} > 0 \\ \sigma_v = u_* \sqrt{0.35\left(\dfrac{h_i}{-kL}\right)^{2/3} + \left(3.6-\dfrac{h}{h_i}\right)} & \text{for } \dfrac{1}{L} > 0 \end{cases}$$

The *vertical* standard deviation $\sigma_z(x)$ is given by

$$\sigma_z = f_{z1}(z_r, L)x^{f_{z2}(L)} \quad \text{(m)} \tag{J.4}$$

The f_{z1} and f_{z2} are functions of the roughness length, z_r and Monin–Obukov length, L. For x in meters, these functions are (with arctan function arguments in radians):

$$f_{z1}(z_r, L) = 0.0053\left\{\arctan\left[C_1\left(\frac{1}{L}-C_2\right)\right]+1.81\right\}$$

$$\left(\frac{C_3}{L}+0.13\right)^{-1.176}\left(\frac{z_r}{C_4}+0.057\right)^{0.285} \quad \left(\text{in m}^{(1-f_{z2})}\right) \tag{J.5}$$

$$f_{z1}(L) = 0.249\left\{\arctan\left[C_5\left(\frac{1}{L}-C_6\right)\right]+1.041\right\} \tag{J.6}$$

The values of the dimensional constants are

$C_1 = 534$ m; $C_2 = 0.0016$ m^{-1}; $C_3 = 1$ m;

$C_4 = 0.10$ m; $C_5 = -214$ m; $C_6 = 0.00077$m^{-1}.

These expressions are valid for $50 < x < 50{,}000$ m, for $-0.12 < 1/L < 0.08$ m^{-1}, and for $0.001 < z_r < 1$ m. For $x < 50$ m, linear interpolation between $\sigma_x = 0$ and $\sigma_x(50)$ should be used.

For elevated release heights, $h \geq h_i$, the $\sigma_z(x)$ function is similar to the $\sigma_y(x)$ function.

For $h \geq 0.1h_i$, the $\sigma_z(x)$ is calculated similar to the calculation of σ_y:

$$\sigma_z = \sigma_w \frac{x}{u_a}\left(1+0.9\sqrt{\frac{x}{u_a t_i}}\right)^{-1} \quad \text{(m)} \tag{J.7}$$

with

$$t_i = \begin{cases} 500 \text{ s}, & 1/L < 0 \\ 100 \text{ s}, & 1/L > 0 \end{cases} \tag{J.8}$$

J.2. Flammable Mass Formulas

The evaluation of the integral, Eq. (5.1.1) requires first finding the dimensions of the cloud to the LFL and UFL.That is, find x_{LFL}, y_{LFL}, and z_{LFL}, representing the cloud length, width, and height.The cloud width and height can be readily found by evaluating the concentration profiles, Eq. (I.1) and (I.2) at the LFL. For example, y_{LFL} is given in terms of the centerline concentration and the standard deviation by

$$y_{\mathrm{LFL}}(x) = \sigma_y(x)\sqrt{2\ln\left(\frac{c(x,0,0)}{c_{\mathrm{LFL}}}\right)} \quad \text{(m)} \tag{J.9}$$

A similar formula applies for z_{LFL}.Approximating x_{LFL} is more difficult. The *Yellow Book* approximate formula for x_{LFL} for a ground-level release is

$$x_{\mathrm{LFL.ground}} = \left\{\frac{q}{\pi 0.3\sigma_{\mathrm{v}} f_{z1}(z_{\mathrm{r}}, L) c_{\mathrm{LFL}}}\right\}^{\frac{1}{f_{z2}(L)+1}} \quad \text{(in m)} \tag{J.10}$$

where the function $f(x, y, z, t)$ is defined as follows:

$$f(x,y,z,t) = \begin{cases} c(x,y,z,t) & \text{if LFL} < c(x,y,z,t) < \text{UFL} \\ 0 & \text{otherwise} \end{cases}$$

For an elevated plume which has not yet touched the ground, the LFL at the centerline is reached at a shorter distance, approximated by:

$$x_{\mathrm{LFL.elevated}} = x_{\mathrm{LFL.ground}}\left\{\frac{1}{2}\right\}^{\frac{1}{f_{z2}(L)+1}} \quad \text{(in m)} \tag{J.11}$$

Plumes which have touched the ground will have a value of x_{LFL} between the elevated and the ground-level values.The plume is considered to touch the ground if $2\sigma_z > h$. Eqations (J.9) and (J.10) apply as well when evaluated at the UFL

To treat a finite-sized source, the *Yellow Book* suggests relating the source dimensions, b_{0y} and b_{0z} to σ s found at the virtual source distances x_{vy} and x_{vz} given by

$$\sigma_y(x_{\mathrm{vy}}) = \frac{b_{0y}}{2.15}, \qquad \sigma_z(x_{\mathrm{vz}}) = \frac{b_{0z}}{2.15} \tag{J.12}$$

For sources for which the virtual source dimensions can be considered equal, $x_{\mathrm{vy}} \approx x_{\mathrm{vz}}, = x_{\mathrm{v}}$, the approximate formulas for the flammable mass of a continuous line source is

$$\frac{f_{\mathrm{m}}}{w}=\frac{1}{u_{\mathrm{a}}}\left\{\frac{f_{z2}(L)+1}{f_{z2}(L)+2}(x_{\mathrm{LFL1}}-x_{\mathrm{UFL}})-\frac{x_{\mathrm{v}}^{(f_{z2}(L)+2)}}{f_{z2}(L)+2}\left(\frac{1}{x_{\mathrm{UFL}}^{(f_{z2}(L)+1)}}-\frac{1}{x_{\mathrm{LFL}}^{(f_{z2}(L)+1)}}\right)\right\}$$

$$\text{if } x_{\mathrm{UFL}}>0 \qquad \text{(J.13)}$$

$$\frac{f_{\mathrm{m}}}{w}=\frac{1}{u_{\mathrm{a}}}\left\{\frac{f_{z2}(L)+1}{f_{z2}(L)+2}(x_{\mathrm{LFL1}}-x_{\mathrm{v}})-\frac{x_{\mathrm{v}}^{(f_{z2}(L)+2)}}{(f_{z2}(L)+2)x_{\mathrm{LFL}}^{(f_{z2}(L)+1)}}\right\} \qquad \text{if } x_{\mathrm{UFL}}<0$$

$f_{\mathrm{m}}/w = 0$ if $x_{\mathrm{LFL}} = 0$, which will be the case when the initial concentration is lower than c_{LFL}.

APPENDIX K

EVALUATING FLAMMABLE MASS FOR GAUSSIAN DISPERSION MODEL—CONTINUOUS RELEASE, RIGOROUS SOLUTION

K.1. Flammable Mass Across Elliptical Cross Section

The flammable mass of a continuous release Gaussian model is given by Hesse (1991) in terms of standard tabulated mathematical functions. Constant concentration isopleths for a Gaussian model are ellipses. The first integration step for a continuous source must be in the *y-z* plane, and in this plane a concentration of value c_0 at a centerline distance x is an ellipse described by:

$$\frac{y^2}{\alpha^2} + \frac{(z-h)^2}{\beta^2} = 1 \tag{K.1}$$

where

$$\alpha = \sigma_y \sqrt{2 \ln\left(\frac{w}{\pi \sigma_y \sigma_z u c_0}\right)} \tag{K.2}$$

$$\beta = \sigma_z \sqrt{2 \ln\left(\frac{w}{\pi \sigma_y \sigma_z u c_0}\right)} \tag{K.3}$$

The integration around an ellipse results in an elliptic integrals of the second kind (or simply elliptic integrals) defined by

$$E(k,\theta) = \int_0^\theta \sqrt{1 - k^2 \sin^2 \phi}\, d\phi, \qquad 0 \le k = 1, \quad 0 \le \theta \le \pi/2$$

with

$$\theta = \begin{cases} \pi/2 & \text{if } h = 0 \\ \arccos\left[\dfrac{\alpha\sqrt{1-h^2/\beta^2}}{\sqrt{\left(h^2 - \alpha^2(1-h^2/\beta^2)\right)}}\right] & \text{if } h > 0 \end{cases} \tag{K.4}$$

The length of perimeter given by Eq. (K.1) in terms of elliptic integrals is

$$p(c,x) = 2\alpha\left[E(k_c, \pi/2) + E(k_c, \theta)\right]$$

where

$$k_c = \frac{\sqrt{\sigma_y^2 - \sigma_z^2}}{\sigma_y} \tag{K.5}$$

Hesse provides different solutions for $h = 0$ and $h > 0$ which are further explained in Appendix I. It suffices to illustrate the nature of the solution here by giving it only for a ground-level release, $h = 0$. A differential volume with a given concentration is expressed by

$$dV_F = p(c,x)\,dy\,dz \tag{K.6}$$

Multiplying this differential volume by its constant concentration, $c(x, y, z)$, gives the flammable mass within a thin elliptical band described by Eq. (K.1). The variable α in Eq. (K.1) is also known as the cloud half-width, or the crosswind distance from the center to the point where the concentration is c. To find the total flammable mass we must integrate the above differential element with respect to y between the half-widths at the flammable limits, $y = \alpha(c_{UFL}, x)$ and $y = \alpha(c_{LFL}, x)$. This gives the cross-sectional area of a slice taken from the cloud perpendicular to the airflow direction with thickness dx located at a centerline distance x from the source. Transforming the variable of integration from y to the corresponding concentration c and then to the corresponding distance x_L gives the total flammable mass as

$$m_F(\text{continuous}) = \int_{\alpha(c_{UFL},x)}^{\alpha(c_{LFL},x)} cp(c,x)\,dy = -\int_{c_{UFL}}^{c_{LFL}} 2\sigma_y^2 E(k_c, \pi/2)\,dc$$

$$m_F(\text{continuous}) = 2(c_{UFL} - c_{LFL}) \int_{x_{UFL}}^{x_{LFL}} \sigma_y^2 E(k_c, \pi/2)\,dx \tag{K.7}$$

To complete the solution, a highly accurate approximating function for the elliptic integral can be substituted into Eq. (K.7) and numerically integrated. Hesse also provides upper and lower bounding solutions which are of comparable complexity but which avoid any numerical integration.

A solution for the puff model is similar, except that it requires a double integration giving

$$m_{\mathrm{F}}(\text{puff}) = \int_{x_{\mathrm{LFL1}}}^{x_{\mathrm{LFL2}}} \int_{a(c_{\mathrm{UFL}},x)}^{a(c_{\mathrm{LFL}},x)} cp(c,x)\,dy\,dz = 2(c_{\mathrm{UFL}} - c_{\mathrm{LFL}}) \int_{x_{\mathrm{LFL1}}}^{x_{\mathrm{LFL2}}} \sigma_{yp}^2 E(k_p, \pi/2)\,dx \quad (\mathrm{K.8})$$

Here k_p is the same as k_c with σ_{yp} replacing σ_y and σ_{zp} replacing σ_z. There is some added complexity, since there are two roots which must be found which form the limits of integration in x, x_{LFL1}, and x_{LFL2}. These arise from evaluating the puff concentration profiles at $y = z = 0$ with sigmas given by Eq. (K.1):

$$c = \frac{w_p}{(2\pi)^{3/2}\, cde(x + x_{\mathrm{v}})^{j+k+l}} \exp\left(\frac{-(x - ut)^2}{2c^2 (x + x_{\mathrm{v}})^{2j}}\right) \quad (\mathrm{K.9})$$

The notation x_{LFL1} refers to the lower root and x_{LFL2} refers to the upper root for the LFL as illustrated in Figure K.1 with LFL= 0.20 for example.

For finding the roots in x at the UFL and LFL, Hesse recommends a simple algorithm such as the bisection method or the Newton-Raphson method described in Cheney and Kincaid (1980).

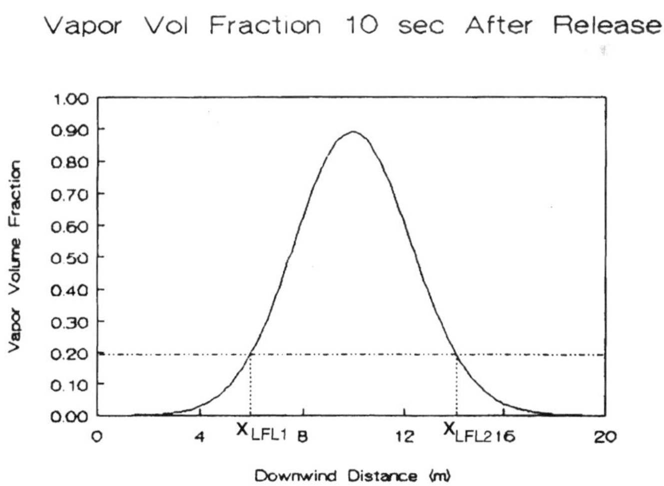

Figure K.1. Example of Gaussian concentration profile indicating the integration limits x_{LFL1} and x_{LFL2}.

K.2. Evaluating Integrals for Ground-Level Continuous and Puff Releases

The analytic solutions in Eq. (K.7) and Eq. (K.9) can be evaluated with approximate analytic functions. To evaluate the integrals involving $E(k, \theta = \pi/2)$, Hesse (1991) recommends using the approximating function given in Abramowitz and Stegun (1970):

$$E(k,\theta) = E(1-\gamma_1) = \sum_{i=0}^{4} a_i \gamma_1^i - \ln(\gamma_1)\sum_{i=0}^{4} b_i \gamma_1^i$$

where

$$\gamma_1 = \begin{cases} \dfrac{b^2 (x+x_v)^{2(n-m)}}{a^2} & \text{for a continuous release} \\ \dfrac{e^2 (x+x_v)^{2(l-k)}}{d^2} & \text{for an instantaneous release} \end{cases} \qquad \text{(K.10)}$$

The coefficients a_i and b_i are given in Table K.1. The error in the approximation of Abramowitz and Stegun is given as

Continuous release: Error $<2\times10^{-8}\left|x_{\mathrm{LFL}} - x_{\mathrm{UFL}}\right|$

Puff release: Error $<2\times10^{-8}\left|x_{\mathrm{LFL1}} - x_{\mathrm{UFL2}}\right|$

The integrals in Eqs. (K.7) and (K.8) can now be readily integrated numerically.

TABLE K.1
Values for Coefficients in Equation (K.8)

i	a_i	b_i
0	0.0	n/a
1	0.44325141463	0.24998368310
2	0.06260601220	0.09200180037
3	0.04757383546	0.04069697526
4	0.01736506451	0.00526449639

K.3. Upper and Lower Bounding Solutions for Ground-Level Clouds

An alternative approach which avoids numerical integration is to find upper and lower bounding solutions. To do this, first change the variables in Eqs. (K.7) and (K.8) to give

$$m_F\,(\text{continuous}) = \frac{a^2(c_{\text{UFL}} - c_{\text{LFL}})}{(n-m)} \int_{Y(c_{\text{UFL}},x)}^{Y(c_{\text{LFL}},x)} \frac{(x+x_v)^{2m+1}}{\gamma_1} E(1-\gamma_1)\,d\gamma_1 \quad \text{(K.11)}$$

and

$$m_F\,(\text{puff}) = \frac{d^2(c_{\text{UFL}} - c_{\text{LFL}})}{(l-k)} \int_{\Omega(c_{\text{UFL1}})}^{\Omega(c_{\text{LFL2}},)} \frac{(x+x_v)^{2k+1}}{\gamma_1} E(1-\gamma_1)\,d\gamma_1$$

with

$$Y(c) = \frac{b^2}{a^2}\left[\frac{w}{\pi abuc}\right]^{\frac{2(n-m)}{(n+m)}}$$
$$\Omega(X) = 1 - \frac{e^2(x+x_v)^{2(l-k)}}{d^2} \quad \text{(K.12)}$$

To transform the integrand of Eqs. (K.11) and (K.12) to a function of γ_1 alone, introduce the variable η which satisfies the equation

$$\gamma_1^{\eta} = (x+x_v)^{2m+1} \quad \text{(K.13)}$$

Solving for η gives

$$\eta = \frac{2m+1}{2(n-m)} \quad \text{(K.14)}$$

So, with this expression for η Eq. (K.11) becomes

$$m_F\,(\text{continuous}) = \frac{a^2(c_{\text{UFL}} - c_{\text{LFL}})}{(m-n)} \int_{Y(c_{\text{LFL}},x)}^{Y(c_{\text{UFL}},x)} \frac{E(1-\gamma_1)}{\gamma_1^{1-\eta}}\,d\gamma_1 \quad \text{(K.15)}$$

and similarly for Eq. (K.12). In this form, there is an analytic solution for integer values of η. For example, using the tabulated values for Stability C, $m = 0.94$ and $n = 0.89$, so $\eta = -28.8$. Rounding values of η to the next integer value and denoting this as s (e.g., $s = -29$) we obtain an upper bounding solution, $UB(m_F)$. Since s is always negative, adding 1 to s will always give a lower bounding solution, $LF(m_F)$ which has the same form except with s replaced by $s + 1$. These bounding solutions are given below.

Grounded continuous release bounding solutions

$$UB(V_{\mathrm{f}}) = \frac{a^{2(s+1)}(c_{\mathrm{UFL}} - c_{\mathrm{LFL}})}{b^{2s}(m-n)} \times \left(\sum_{i=0}^{4} \frac{a_i}{s+i} \gamma_1^{s+i} + \sum_{i=0}^{4} \frac{b_i}{(s+i)^2} \gamma_1^{s+i} - \ln(\gamma_1) \sum_{i=0}^{4} \frac{b_1}{s+i} \gamma_1^{s+i} \right) - 2 \times 10^{-8} \gamma_1 \Bigg|_{\gamma_1 = Y(c_{\mathrm{LFL}})}^{\gamma_1 = Y(c_{\mathrm{UFL}})} \tag{K.16}$$

$$LB(V_{\mathrm{f}}) = \frac{a^{2(s+1)}(c_{\mathrm{UFL}} - c_{\mathrm{LFL}})}{b^{2s}(m-n)} \times \left(\sum_{i=0}^{4} \frac{a_i}{s+i+1} \gamma_1^{s+i+1} + \sum_{i=0}^{4} \frac{b_i}{(s+i+1)^2} \gamma_1^{s+i+1} - \ln(\gamma_1) + \sum_{i=0}^{4} \frac{b_1}{s+i} \gamma_1^{s+i+1} \right) - 2 \times 10^{-8} \gamma_1 \Bigg|_{\gamma_1 = Y(c_{\mathrm{LFL}})}^{\gamma_1 = Y(c_{\mathrm{UFL}})} \tag{K.17}$$

Grounded puff release bounding solutions

$$UB(V_{\mathrm{f}}) = \frac{d^{2(v+1)}(c_{\mathrm{UFL}} - c_{\mathrm{LFL}})}{e^{2v}(k-l)} \times \left(\sum_{i=0}^{4} \frac{a_i}{v+i} \gamma_1^{v+i} + \sum_{i=0}^{4} \frac{b_i}{(v+i)^2} \gamma_1^{v+i} - \ln(\gamma_1) \sum_{i=0}^{4} \frac{b_1}{v+i} \gamma_1^{v+i} \right) - 2 \times 10^{-8} \gamma_1 \Bigg|_{\gamma_1 = \Omega(c_{\mathrm{UFL}}^2)}^{\gamma_1 = \Omega(c_{\mathrm{UFL}}^1)} \tag{K.18}$$

$$LB(V_{\mathrm{f}}) = \frac{d^{2(v+1)}(c_{\mathrm{UFL}} - c_{\mathrm{LFL}})}{e^{2v}(k-l)} \times \left(\sum_{i=0}^{4} \frac{a_i}{v+i+1} \gamma_1^{v+i+1} + \sum_{i=0}^{4} \frac{b_i}{(v+i+1)^2} \gamma_1^{v+i+1} - \ln(\gamma_1) \sum_{i=0}^{4} \frac{b_1}{v+i+1} \gamma_1^{v+i+1} \right) - 2 \times 10^{-8} \gamma_1 \Bigg|_{\gamma_1 = \Omega(c_{\mathrm{UFL}}^2)}^{\gamma_1 = \Omega(c_{\mathrm{UFL}}^1)} \tag{K.19}$$

K.4. Elevated Continuous Releases

To generalize the ground-level release solutions, recall that the surface area occupied by the isopleths is given by $A_F(c)$ defined as

$$A_F(c) = \int_0^{x_L(c)} p(c, \omega)\, d\omega \tag{K.20}$$

The flammable mass is given by integrating this area times the mass concentration giving:

$$m_F(c) = \int_{c_{LFL}}^{c_{LFL}} c A_f(c)\, dx_L(c) \tag{K.21}$$

This result is a triple integral for which we would have to resort immediately to numerical methods in order to evaluate it directly. Reversing the order of the second and third integrals followed by an application of a change of variables gives an equivalent form, namely

$$m_F = -\left(\frac{w}{\pi abu}\right)^{\mu} \int_{x_{LFL}}^{x_{LFL}} \int_{c_{LFL}}^{\xi} \frac{2\sqrt{2}\sigma_y [E(k_c, \pi/2) + E(k_c, \theta)]}{m+n} c^{\mu} \sqrt{\ln\left(\frac{\xi}{c}\right)}\, dc\, dx$$

with

$$\mu = \frac{1-m-n}{m+n}$$
$$\xi = \frac{w}{\pi \sigma_y \sigma_z u} \tag{K.22}$$

Part of the integrand is the improper integral

$$\Gamma = \int c^{\mu} \sqrt{\ln\left(\frac{\xi}{c}\right)}\, dc \tag{K.23}$$

Integrating by parts and using induction, the expression for Γ can be evaluated by its series representation

$$\Gamma = -\frac{2}{3} c^{\mu+1} \left[\ln\left(\frac{\xi}{c}\right)\right]^{3/2} \sum_{i=0}^{\infty} \frac{p^i}{(c)_i}$$

where

$$p = (\mu+1) \ln\left(\frac{\xi}{c}\right)$$
$$c = 5/2, \quad (c)_i = c(c+1)(c+2)\cdots(c+i) \tag{K.24}$$

The above infinite series is known as the confluent hypergeometric function, represented by ${}_1F_1[1; c; w]$. and fully described by Magnus, Oberrhettinger, and Soni (1966). This function converges for all z provided $c \neq -m$ where $m = 0, 1, 2, \ldots$. The final solution for an elevated continuous release is

$$m_{\mathrm{F}} = -\left(\frac{w}{\pi abu}\right)^{\mu} \int_{x_{\mathrm{UFL}}}^{x_{\mathrm{LFL}}} \left\{ \Psi E(k_c, \pi/2) + \int_{c_{\mathrm{LFL}}}^{\xi} \frac{2\sqrt{2}\sigma_y E(k_c, \theta)}{m+n} c^{\mu} \sqrt{\ln\left(\frac{\xi}{c}\right)}\, dc \right\} dx \tag{K.25}$$

with

$$\Psi = \frac{2\sqrt{2}\sigma_y}{3(m+n)} \left[\frac{w}{\pi abu}\right]^{\mu} c_{\mathrm{LFL}}^{\mu+1} \left[\ln\left(\frac{\xi}{c_{\mathrm{LFL}}}\right)\right]^{3/2} \left\{ {}_1F_1\left[1; \frac{5}{2}; (\mu+1)\ln\left(\frac{\xi}{c_{\mathrm{LFL}}}\right)\right]\right\} \tag{K.26}$$

Press et al. (1989) provide algorithms for computing $E(k_c, \theta)$ for $0 \leq \theta \leq \pi/2$.

APPENDIX L

NUMERICAL INTEGRATION TO FIND FLAMMABLE MASS

L.1. Treatment of Generalized Concentration Profiles

The models PHAST/SAFETI and DRIFT discussed in Section 5.2 make use of generalized concentration profiles. The assumed form for the concentration profiles initially are sharp-edged and gradually diffuse to Gaussian form or slightly more diffuse than Gaussian. Numerical integration applied to this form, therefore, also applies to Gaussian models as a special case.

L.2. Limits for Integration

The flammable mass is found by numerical integration between the flammable limits. The distance to flammable limits is found by inverting Eqs. (5.2.40), (5.2.41), and (5.2.43). For example, the vertical distance to the upper LFL contour is

$$z_{\mathrm{LFL,\,max}} = a_3 \ln\left[\left(\frac{c_0(x)}{c_{\mathrm{LFL}}}\right)^{1/s}\right] \tag{L.1}$$

The vertical distance to the lower LFL contour is similarly

$$z_{\mathrm{LFL,\,min}} = \min\left\{\left|\frac{z_0}{\cos\theta}\right|, z_{\mathrm{LFL}}\right\} \tag{L.2}$$

The integration in z is done in an outer loop and the integration in y in an inner loop, each by a modified Simpson's rule method given by Press et al. (1986). First a rough approximation is made by using a single step across the limits of integration. Then the number of steps is doubled repeatedly until the change in the integration result falls within acceptable limits. For example, the first step in the z integration uses the result of the integration in y, F_{yint}, and finds

$$F_{z\,\min} = \exp\left[-\left|\frac{z_{\mathrm{LFL,\,min}}}{a_3}\right|^s\right] F_{y\,\mathrm{int}}$$

$$F_{z\,\max} = \exp\left[-\left|\frac{z_{\mathrm{LFL,\,max}}}{a_3}\right|^s\right] F_{y\,\mathrm{int}}$$

$$F_{\mathrm{M}} \cong 0.5(z_{\mathrm{LFL,\,max}} - z_{\mathrm{LFL,\,min}})(F_{z\,\min} + F_{z\,\min}) \tag{L.3}$$

The numerical integration can be verified against analytical integration over the entire mass of a plume. The analytical integration is done over the limits from the ground to infinity as follows:

$$A(x) = 2 \int_{-h/\cos\theta}^{\infty} \int_{0}^{\infty} \exp\left[-\left(\frac{z}{a_3}\right)^s\right] \exp\left[-\left(\frac{y}{a_2}\right)^m\right] dy\,dz \tag{L.4}$$

The resulting solution is abbreviated in terms of the effective half-height, H_{eff}, the effective half-width, W_{eff}, and a term, h_{d}, ranging in value from 1 for a plume aloft (elliptical cross-section) to 0 for a grounded plume (semi-ellipse cross-section):

$$A = 2H_{\mathrm{eff}} W_{\mathrm{eff}} (1 + h_{\mathrm{d}}) \tag{L.5}$$

$$H_{\mathrm{eff}} = \frac{1}{c(x,y,0)} \int_0^{\infty} c(x,y,z)\,dz = \int_0^{\infty} F_z(x,z)\,dz = \Gamma\left(1 + \frac{1}{s}\right) a_3(x) \tag{L.6}$$

$$W_{\mathrm{eff}} = \frac{1}{c(x,0,z)} \int_0^{\infty} c(x,y,z)\,dy = \int_0^{\infty} F_z(x,y)\,dy = \Gamma\left(1 + \frac{1}{m}\right) a_2(x) \tag{L.7}$$

$$h_{\mathrm{d}} = P\left[\frac{1}{s}, \left(\frac{h/\cos\theta}{a_3}\right)^s\right] \tag{L.8}$$

where the gamma function $\Gamma(z)$ is defined by

$$\Gamma(z) = \int_0^{\infty} t^{z-1} e^{-t}\,dt \tag{L.9}$$

and the partial gamma function $P(a,b)$ is defined by

$$P(a,b)=\frac{1}{\Gamma(a)}\int_0^b t^{a-1}e^{-t}\,dt \tag{L.10}$$

For the special case when $s = m = 2$, the Gaussian case, the results reduce to

$$h_\mathrm{d} = \mathrm{erf}\left(\frac{h}{a_3 \cos\theta}\right) \tag{L.11}$$

$$A(x)=\tfrac{1}{2}(1+h_\mathrm{d})\pi a_2 a_2 \tag{L.12}$$

APPENDIX M

EXPANSION VELOCITY AND DISCHARGE COEFFICIENTS

M.1. Expansion Velocity

The expansion velocity is the final velocity attained after a compressible (gas, vapor, or two-phase) material discharged under choking flow conditions expands from the choked pressure to atmospheric pressure. Taking for example a horizontal discharge from a nozzle or orifice, a vertical plane at the discharge point is taken as Plane 1. The discharge converges to the *vena contracta* at the choke point. A vertical plane perpendicular to the flow at the choke point is Plane 2. The flow becomes supersonic and expands until the pressures reach ambient conditions at a vertical Plane 3. Air entrainment is assumed to begin after the flow reaches the expansion velocity at Plane 3.

The area at the choke point, A_2, is related to the area at the discharge point, A_1 by the discharge coefficient, C_D by:

$$A_2 = C_D A_1 \tag{M.1}$$

A force balance attributed to Dryden et al. (1936), including the annular area, $A_3 - A_1$ gives

$$G_1 A_1 u_1 - G_3 A_3 u_3 + P_1 A_1 + P_3 (A_3 - A_1) - P_3 A_3 = 0 \tag{M.2}$$

which, since

$$w = G_1 A_1 = G_3 A_3 \tag{M.3}$$

reduces to the formula for expansion velocity, u_3:

$$u_3 = u_1 + \frac{P_1 - P_3}{G_1} \tag{M.4}$$

where the mass flux at the choke point, G_2 is the readily evaluated quantity, and:

$$G_1 = C_D G_2 \tag{M.5}$$

M.2. Discharge Coefficients

The formulas given by Jobson (1955) for the discharge coefficient for two-phase flow are obtained from the single-phase discharge coefficients for gas flow, C_{Dg}, and for liquid flow, C_{DL} which are defined as the ratio of the cross-sectional areas occupied by each phase, A_g and A_L to the total area, A_T:

$$C_{Dg} = \frac{A_g}{A_T}, \qquad C_{DL} = \frac{A_L}{A_T}$$

The two-phase discharge coefficient follows by substituting for the phase area terms as a function of the phase specific volumes, v_g and v_L, the slip velocity $K = u_g/u_L$, and the mass fraction of vapor x:

$$C_D = \frac{A_g + A_L}{A_T} = \frac{xv_g + K(1-x)v_L}{\dfrac{xv_g}{C_{Dg}} + \dfrac{K(1-x)v_L}{C_{DL}}} \tag{M.6}$$

For homogeneous equilibrium, $K = 1$. Otherwise, there are formulas for the slip velocity, K (see Chisholm,1987). The gas-phase discharge coefficient is given in terms of the pressure ratio $\eta = P/P_0$, the choked pressure ratio given by Eq. (4.7) and the dimensionless choked mass flux, G^{*2}_{cg} , given by Eq. (4.8) only without the discharge coefficient term [C_D = 1 in Eq. (4.8)]:

$$C_{Dg} = \frac{1}{2f\eta_c^{1/\kappa}}\left[a_1 - \left\{a_1^2 - \frac{4\eta_c^{2/\kappa}(1-\eta)f}{G^{*2}_{cg}}\right\}^{1/2}\right] \tag{M.7}$$

where f is a constant value since C_{DL}= 0.61:

$$f = \frac{1}{C_{DL}} - \frac{1}{2C_{DL}^2} \tag{M.8}$$

and

$$a_1 = \frac{\eta_c - \eta}{G^{*2}_{cg}}\eta_c^{1/\gamma} \tag{M.9}$$

For the case of unchoked flow, $\eta > \eta_c$, then set $\eta_c = \eta$ and a_1 becomes unity.

Again γ is the heat capacity ratio, which is valid with orifice flow since there is too little time for sufficient heat transfer to allow equilibrium to exist. When the phases maintain equal temperatures a generalization of the isentropic constant is preferable (see Chisholm, 1983, p. 207):

$$\gamma = \frac{(1-x)C_{pL} + xC_{pg}}{(1-x)C_{pL} + xC_{vg}} \tag{M.10}$$

APPENDIX N

CONVERSION FACTORS

To convert	From SI Units	To US Units	Multiply by	And add
Area	m^2	ft^2	10.764	
Concentration	kg of i/m^3	lb.m of i/ft.^{3}	1./ 16.0185	
Density	kg/m^3	lb.m/ft$^{\cdot 3}$	1. /16.0185	
Drop diameter	mm	inches	10^{-4}/2.54	
Energy	J = N.m	BTU	947.81	
Flow rate (mass)	kg/s	lb.m/s	2.2048	
Force	N	lb.f	0.22481	
Heat capacity	kJ/(kg.K)	BTU/(lb.F)	0.238846	
Length	m	ft.	3.2808	
Mass	kg	lb.m	2.2048	
Power	W = J/s	BTU/hr	0.26328	
Pressure	Pa = N/m^2	psi	14.696/101325.	
Speed	m/s	ft/s	3.2808	
Surface tension	Pa.s = N/m	lb.f/ft	0.73756	
Temperature	K	R	1.8	
Temperature	K	F	1.8	–459.67
Volume	m^3	ft.^{3}	35.3147	
Volume	m^3	gallon	264.18	

REFERENCES

Abramowitz, M. and Stegun, I. A. *Handbook of Mathematical Functions with Formulas, Graphs, and Mathematical Tables,* 9th printing, U.S. Dept. of Commerce, National Bureau of Standards, Applied Mathematics Series 55, U. S. Government Printing Office, Washington, DC, Nov. 1970.

Alix, P, C. Kobert and Bigot, J-P., "Loss of containment—experimental validation of two-phase critical flow models against water, R11, and methanol," *9th International Symposium on Loss Prevention and Safety Promotion in the Process Industries*, Barcelona, Spain, May 4–7, 1998, pp. 976–986.

American Institute of Chemical Engineers (AIChE), *Dow's Fire and Explosion Index Hazard Classification Guide,* 7th ed. American Institute of Chemical Engineers, New York, 1994.

AIChE, *Dow's Chemical Exposure Index Guide,* 1st ed. American Institute of Chemical Engineers, New York, 1994.

American Petroleum Institute (API), "Surface Roughness Effects on Heavier-Than-Air Gas Diffusion," API Publication No. 4459, Oct. 1987.

Andrews, J., R. Smith, and Gregory, J., "Procedure to calculate the explosion frequency for a module on an offshore platform," Trans. I Chem. E., **72**, Part B, pp. 69–82, 1994.

API, "Effect of Homogeneous and Heterogeneous Surface Roughness on Heavier-Than-Air Gas Dispersion," API Publication No. 4491, March 1989.

API, "Hazard response modeling uncertainty (a quantitative method)," Vol. II, "Evaluation of commonly used hazardous gas dispersion models," API Publication No. 4546, Oct.1992.

ASTM (American Society for Testing and Materials), "Significance of Tests for Petroleum Products," Special Technical Publication 7C, K. Boldt and B. R. Hall, eds., West Conshohocken, PA, 1977, pp. 35–36.

Auer, A. H. Jr., "Correlation of land use and cover with meteorological anomalies," *J. Applied Meteorology,* **17**, 636–643, 1978.

Baines, W. D., and Turner, J. S., "Turbulent buoyant convection from a source in a confined region," *J. Fluid Mech.,* **37**, 51–80, 1967.

Baker, Q. A., Doolittle, C. M., Fitzgerald, G. A., and Tang, M. J., "Recent developments in the Baker–Strehlow VCE analysis methodology," *31st Loss Prevention Symposium*, Houston, TX, AIChE, March 9–13, 1997, Paper 42f.

Baker, Q. A., Tang, M. J., Scheier, E., and Silva, G. J., "Vapor cloud explosion analysis," *AIChE 28th Annual Loss Prevention Symposium,* Houston, TX, April 17–21, 1994.

Baker, W. E., Cox, P. A., Westine, P. S., Culesz, J. J., and Strelow, R. A., *Explosion Hazards and Evaluation.* Elsevier,. Amsterdam, 1983.

Baker, W. E., "Explosions in Air," Wilfred Baker Engineering, San Antonio, TX, 1973, pp. 54–64, 154.

Bartknecht, W., *Explosionschutz: Grundlagen und Anwendung,* Springer-Verlag, Berlin, 1993.

Birch, A. D., Brown, D. R., and Dodson, M. G., Eighteenth Symposium (International) on Combustion, The Combustion Institute, 1981.

Birch, A. D., Brown, D. R. Dodson, M.G., and Swaffield, F., "The structure and concentration decay of high pressure jets of natural gas," *Combust. Sci., and Tech.,* **36**, 249, 1984.

Birch, A. D., Hughes, D. J., and Swaffield, F., "Velocity decay of high pressure jets," *Combust. Sci., and Tech.,* **52**, 161–171, 1987.

Birch, A. D., Brown, D. R., Fairweather, M., and Hargrave, G. K., "An experimental study of a turbulent natural gas jet in a cross-flow," *Combust. Sci., and Tech.,* **66**, 217–232, 1989.

Birtwisle, J., Personal communication, 1999.

Bjerketvedt, D., Bakke, J. R., and van Wingerden, K., "Gas Explosion Handbook." Christian Michelsen Research. CMR-93-A25034, 1993.

Blockley, D. I., and Henderson, J. R., *Proc. Inst. Civ. Eng.,* Part 1, **68**, 719, Nov. 1980.

Bolk, J. W., "Influence of Hydrodynamics on the Upper Explosion Limit of Ethene–Air–Nitrogen Mixtures," Ph.D. Dissertation, The University of Twente, The Netherlands, 1998.

Bond, J., *Flammability Characteristics of Chemicals and Products,* Butterworth-Heinemann, London, 1991, pp. 71–127.

Bowen, P. J., and Shirvill, L. C., "Combustion hazaards posed by the pressurized atomization of high-flashpoint liquids," *J. Loss Prev. Process Ind.,* **7** (3) 233–241, 1994.

Bragg, S. L., "Effect of compressibility on the discharge coefficient of orifices and convergent nozzles," J. Mech. Eng. Sci., **2**, 35–44, 1960.

Britter, R. E., "Special Topics on Dispersion of Dense Gases," Report on Contract Number 1200/01.01, Research and Laboratory Services Div., Health & Safety Executive, Sheffield, UK, 1982.

Britter, R. E., and McQuaid, J., "Workbook on the Dispersion of Dense Gasses," HSE Contract Research Report No. 17/1988, Health & Safety Executive, Sheffield, UK, 1988.

Britton, L. G., "Measurement of ignition energy for safety application," *AIChE Loss Prevention Symposium No. 25*, 1991.

Brown, R., York, J. L., "Sprays formed by flashing liquid jets," *AIChE Journal*, **8**, 149–153, 1962

Bruderer, R. E., "Ignition capabilities of hot surfaces and mechanically generated sparks in flammable gas and dust/air mixtures," Paper 51d, AIChE Summer National Meeting, Minneapolis, MN, August 16–19, 1987.

Burgoyne, J. H., "The flammability of mists and sprays," *Second Symposium on Chemical Process Hazards,* I Chem E.5, 1963, pp 1–.

Bushnell, D. M., and Gooderum, P. B., "Atomization of superheated water jets at low ambient temperatures," *J. Spacecraft & Rockets*, **5**, 231–232, February 1968.

Businger J.A., Wyngaard, J.C., Izumi, Y. and Bradley, E.F., "Flux–Profile Relationships in the Atmosphere Surface Layer," *J Atmos Sci,* **28**, 181–189, March 1971.

Cardillo, P., "Flammability of gas–powder hybrid systems," *Igegneria della Sicurezza e della Prevenzione,* **11**, (1), 12–22, Jan. 1980.

Caulfield, M., Cleaver, R. P. Cook, D. K., and Fairweather, M., "An integral model of turbulent jets in a cross-flow. Part 1-gas dispersion," *Trans IChemE,* **71,** Part B, 235–242, 1993.

CCPS (Center for Chemical Process Safety), *Concentration Fluctuations and Averaging Time in Vapor Clouds,* AIChE. New York, 1995.

CCPS, *Guidelines for Chemical Process Quantitative Risk Analysis,* AIChE, New York, 1989.

CCPS, *Guidelines for Chemical Transportation Risk Analysis,* AIChE, New York, ISBN 0-8169-0629-7, 382 pp., 1995.

CCPS. *Guidelines for Evaluating the Characteristics of Vapor Cloud Explosions, Flash Fires, and BLEVEs,* AIChE, New York., New York, ISBN 0-8169-0474-X, 387 pp., 1994.

CCPS, *Guidelines for Evaluating Process Plant Buildings for External Explosions and Fires,* AIChE. New York, 1996.

CCPS, *Guidelines for Hazard Evaluation Procedures, 2nd Edition with Worked Examples,* AIChE, New York, 1992.

CCPS, *Guidelines for Investigating Chemical Process Incidents,* AIChE, New York, 1992.

CCPS, *Guidelines for Use of Vapor Cloud Dispersion Models,* 2nd ed., AIChE, New York, 1996.

CCPS, *Guidelines for Postrelease Mitigation Technology,* AIChE, New York, 1997.

CCPS, *RELEASE: A Model with Data to Predict Aerosol Rainout in Accidental Releases,* AIChE, New York, 1998.

CCPS, *Understanding Atmospheric Dispersion of Accidental Releases,* AIChE, New York, 1995.

CERC (Cambridge Environmental Research Consultants), "GASTAR dense gas dispersion model version 2.2 user's guide," Cambridge, UK, 1990.

Chan, S. T., "FEM3A—A finite element model for the simulation of gas transport and dispersion: User's manual," LLNL report UCRL-21043, April 1988.

Checkel, M. D., Ting, D. S. K., and Bushe, W. K., *J., Loss Prev. Process Ind.,* **8,** 215, 1995.

Cheney, W., and Kincaid, D., *Numerical Mathematics and Computation,* Brooks/Cole Publ. Co., 1980, pp. 41–53.

Chisholm, D., *Two-Phase Flow in Pipelines and Heat Exchangers,* George Godwin, Longman House, Burnt Mill, Harlow, Essex, England, 1983, p. 205.

Claessen, G. A., Vliegen, E., J. G., Joosten, G. E. H., and Geersen, T. M., "Flammability characteristics of natural gases in air at elevated pressures and temperatures," *Proc. 5th Int. Symp. Loss Prevention and Safety Promotion in the Process Industries,* Cannes, France, 1986.

Clay, G.A., Fitzpatrick, R. D., Hurst, N. W., Carter, D. A., and Crossthwaite, P. J., "Risk assessment for installations where liquefied petroleum gas (LPG) is stored in bulk vessels above the ground," *J. Hazardous Materials,* **20**, 1988.

Cleaver, R. P., Marshall, M. R., and Linden, P. F., "The build-up of concentration within a single enclosed volume following a release of natural gas," J. *Hazardous Materials,* **36**, 209–226, 1994.

Cleaver, R. P., Cooper, M. G., and Halford, A. R., "Further developments of a model for dense gas dispersion over real terrain," *J. Hazardous Materials,* **40**, 85–108, 1995.

CMR (Christian Michelsen Research AS), "Gas Explosion Handbook, Version 1.2," Ref. No. CMR-93-A25034, p. 63, Bergen, Norway, 1993. (Also listed as Bjerketvedt, 1993.)

Colenbrander G. W., and Puttock J. S., "Maplin Sands experiments 1980: interpretation and modeling of liquefied gas spills on sea" in (Ooms G., and Tennessee H., Eds.) *Atmospheric Dispersion of Heavy Gases and Small Particles*, Springer-Verlag, New York, 1984.

Cook, S. J., Cullis, C. F., and Good, A. J., "The measurement of the flammablililty limits of mists," *Combustion and Flame,* **30**, 309–317, 1977.

Considine, M., Grint, G. C., and Holden, P. L., "Bulk storage of LPG- factors affecting offsite risk," *I. Chem. E. Symp. Series,* No. 71, The Assessment of Major Hazards, 1982.

Coward, H. F., and Jones, G. W., "Limits of flammability of gases and vapors," *Bur. of Mines Bulletin* 503, 1952.

Cox, A., Lees, F., and Ang, M., "Classification of Hazardous Locations," a report of the Inter-Institutional Group Classification of Hazardous Locations (IIGCHL), 1990.

Crank, J., *The Mathematics of Diffusion,* Oxford U. P:ress, London, 1956.

CRC Press, *CRC Handbook of Chemistry and Physics,* Boca Raton, FL, 1997.

Crowl, D. A., and Louvar, J. F., *Chemical Process Safety: Fundamentals with Applications,* Prentice Hall, Englewood Cliffs, NJ, 1990, pp. 163–164, p. 373.

Cullen, The Honorable Lord, "The Public Inquiry into the Piper Alpha Disaster," HM Stationery Office, London, 1990.

Csanady, G. T., *Atmos. Env.,* **3**, 25–46, 1969.

Darby, R., "Viscous two-phase flow through safety relief valves engineering design report," 20th DIERS Users Group Meeting, Scottsdale, Arizona, April 14–16, 1997.

Daubert, T. E., and Danner, R. P., *Physical and Thermodynamic Properties of Pure Chemicals, Data Compilation,* AIChE, New York, 1989.

Davenport, J. A., "A survey of vapor cloud incidents," *Chem. Eng. Progress,* **73**(9), September 1977. Also *AIChE Loss Prevention,* **11**, p. 39, 1977.

Davenport, J. A., "A study of vapor cloud incidents—an update," I. Chem. E. Symposium Series No. 80, *4th Int. Symp. on Loss Prevention and Safety Promotion in the Process Industries,* **1**, 1983.

DeMarrais, G.A., "Wind speed profiles at Brookhaven National Laboratory," *J Met*, **16**, 181–190, 1959.

Department of Employment (UK), "The Flixborough Disaster; Report of the Court of Inquiry," Her Majesty's Stationery Office, 1995.

DeVaull, G. E., and King, J. A., "Similarity scaling of droplet evaporation and liquid rainout following the release of superheated flashing liquid to the environment" *85th Annual Meeting, Air & Waste Management Assoc.,* Kansas City, MO, June 21–26, 1992.

Dodge, F., Bowels, E., and White, R., "Release rate of hazardous chemicals from damaged cargo vessels," *Proc. 1980 National Conference on Control of Hazardous Materials Spills,* 1980.

Donaldson, C., and Snedecker, C., "A study of free jet impingement," *J. Fluid Mechanics,* 45, 281–319, 1971.

DNV Technica, Inc., *SAFETI User's Manual,* London and Houston, 1988.

DNV Technica, Inc., *PHAST Version 2.0 Theory Manual,* London and Houston, 1989.

DNV Technica, Inc., *Consequence Modeling Documentation, SAFETI Version 3.41,* Houston, October 1997.

Drivas, P. J., "Calculation of evaporative emissions from multicomponent liquid spills," *Environ. Sci. Technol.*, **16**, 726–728, 1982.

Dryden, H. L., Murnaghan, F. D., and Bateman, H., in *Hydrodynamics* Dover Publications, New York, 1936, p. 540.

Duijm, N. J., Jones, S. J., Martin, D., and Weber, D. M., "The effect of obstacles on dense gas clouds," in (J. J. Mewis, H. J. Pasmas, and E. E. DeRademaeker, Eds.), *Loss Prevention and Safety Promotion in the Process Industries,* **1**, Elsevier Science Publishers, New York, 1995, pp. 185–196.

Dumas, R., Arzoumanian, E., and Fulachier, L., "Probabilities conditionmelles spatiotemporelles des fluctuations de temperature dans un couche limite turbulente," *Comptes Rendues Acad. Sci. Paris*, **284B**, 487–490, 1977.

Dyer, A. J., and Hicks, B. B., "Flux-gradient relationships in the constant flux layer," Quart. *J. Roy. Meteorol. Soc.,* **96**, 715–721, 1970.

Edgerton, A., and Powling, J., "The limits of flame propagation at atmospheric pressure. II. The influences of changes in the physical properties," *Proc. Roy. Soc.,* **193A**, 1948.

Eggen, J. B. M. M., "Guidance for Application of the Multi-Energy method," TNO Prins Maurits Laboratory report, March, 1995.

Eichhorn, J, "Careful! Mists can explode," *Petroleum Refiner,* **34** (11) 194–196, Nov. 1955.

Ely, J. F., and M. L. Huber, "NIST Thermophysical Properties of Hydrocarbon Mixtures Database, Version 1.0 User's Guide," US Dept. of Commerce, National Institute of Standards and Technology, Standard Reference Data Program, Gaithersburg, MD 20899, Feb. 1990.

EPA (U. S. Environmental Protection Agency) "Model risk management program and plan for ammonia refrigeration," by Science Applications International Corp., Table 2.1, (May 1996a).

EPA (U. S. Environmental Protection Agency) "RMP offsite consequence analysis guidance" (May 24, 1996b).

Ermak, D. L., and Chan, S. T., "Recent developments on the FEM3 and SLAB atmospheric dispersion models," *Proc. IMA Conf. on Stability Sratified Flows and Dense Gas Dispersion,* Chester, UK, April 9–10, 1986.

Ermak, D. L., "User's Manual for SLAB: At Atmospheric Dispersion Model for Denser-Than-Air Releases," Lawrence Livermore National Laboratory, Livermore, CA 94550, 1989.

Factory Mutual, "Properties of flammable liquids, gases, and solids," *Ind. Eng. Chem.*, **32,** 880–884, 1940.

Factory Mutual Engineering Corp., *Handbook of Industrial Loss Prevention,* McGraw-Hill, New York, 1967.

Fauske, H.K., "The Discharge of Saturated Water Through Tubes," *Chem. Eng. Prog. Symp. Series,* **61**, 210–216, 1965.

Fauske, H. K., "Flashing flows or some practical guidelines for emergency releases," *Plant/Operations Progress,* **4**(3) 132–134, July 1985.

Fay, J. A., and Lewis, D. H., The inflammability and dispersion of LNG vapor clouds, *Proc. 4th Int. Symp. on Transport of Hazardous Cargoes by Sea and Inland Waterway,* U. S. Coast Guard, Washington, DC, 1975, pp. 489–498.

Fiock, E. F., "Measurement of burning velocity," in (B. Lewis, R. N. Pease, and H. S. Taylor, Eds.) *High Speed Aerodynamic and Jet Propulsion,* Vol. IX, pt. 2. Geoffry Cumberledge/Oxford Univ. Press, London, 1955, p. 409.

Flacher, A., "Experfimentelle untersuchung der ausbreitung einer schwergaswolke auf geneigtem grund," Thesis supervised by J. Muller and T. K. Fannelop, Institut für Fluiddynamik, ETH Zurich, Switzerland, 1994.

Frie, J. L., Page, G. A., and Buckland, A. C., "Improved evaporation model for spills of multicomponent nonideal liquids," AIChE Symposium Series No. 288, 88, *Heat Transfer Conf., San Diego, CA,* 1992, pp. 344–349.

Gaydon, A. G., and Wolfhard, H. G., *Flames: Their Structure, Radiation, and Temperature,* Chapman and Hill, London, 1970.

Geisbrecht, H., et al., "Explosion hazard analysis of inflammable gas released spontaneously into the atmosphere," *Chem. Ing. Tech.*, **52** (2) 114, 1980. [in German]

Gerstein, M., Levine, L., and Wong, E. L., "Fundamental flame velocities of pure hydrocarbons," NACA Research Memorandum, RM E50G24, National Advisory Committee for Aeronautics, Washington, DC, 1950.

Gibbon, H. J., Wainwright, J., and Rogers, R. L., "Experimental determination of flammability limits of solvents at elevated temperatures and pressures," *Hazards,* **XII,** p. 1, 1994.

Gifford, F. A., "Turbulent diffusion typing schemes: A review," under (R. L. Schoup, Ed.) "Consequences of Effluent Release," *Nucl. Safety* **17** (1), 68–86, 1976.

Glasstone, S., "The Effects of Nuclear Weapons," U.S. Atomic Energy Commission, p. 147, April, 1961, reprinted by the U. S. Dept. of Defense, Washington, DC, 1994.

Glickman, T. S., and Ujihara, A. M., "Deciding between in-place protection and evacuation in toxic vapor cloud emergencies," *J. Hazardous Materials,* **23**, 57–72, 1990.

Green, R. L., and Dresser, D. E., "Heat transfer fluid fires and their prevention in vapor thermal liquid systems," AIChE National Meeting, Paper 9d, Houston, TX, April 4, 1989.

Gresho, P. M., Chan, S. T., Upson, C. D., and Lee, R. L., *Int. J. Num. Meth. Fluids,* **4**, 557, 619, 1984;

Grolmes, M. A., Leung, J. C., and Fauske, H. K., "A review of simplified multi-phase flow methods for emergency relief calculations," Paper No. 4a, AIChE Meeting, Orlando, Florida, March 18–22, 1990.

Grot, R. A., and Persily, A. K., "Measured air infiltration and ventilation rates in eight large office buildings," *Am. Soc. for Testing Materials,* 151–183, 1986.

Gugan, K., "The explosion and fire at New Yorkpro (UK) Ltd., Flixborough," Flixborough Court of Inquiry, J. H. Burgoyne and Partners, London, 1975.

Gugan, K., "Flixborough—a combustion specialist's viewpoint," *Chem. Engr., London,* **309**, 341, 1976.

Gugan, K., *Unconfined Vapor Cloud Explosions,* IChemE, London, 1979.

Gustafson, R. M, and Mudan, K.S., "Ignition potential distribution for heavy gas plumes," *Intl. Conf. on Vapor Cloud Modeling, Boston, Mass.*, 816–843. Nov. 2–4, 1987.

Hanna, S. R., Briggs, G. A., and Hosker, R. P., *Handbook on Atmospheric Diffusion,* DoE/TIC-11223, 102 pp., 1982.

Hanna, S. R., "Applications in air pollution modeling," Chapter 7 in *Atmospheric Turbulence and Air Pollution Modeling,* D. Reidel, Dordrecht, The Netherlands, 1982, pp. 275–310.

Hanna, S. R., Chang, J. C., and Strimaitis, D. G., "Hazardous gas model evaluation with field observations," *Atmospheric Environment,* **27A**(15) 2265–2285, 1993.

Hansel, J. G., Mitchell, J. W., and Klotz, H. C., "Predicting and controlling flammability of multiple fuel and multiple inert mixtures," *Plant/Operations Prog.* **11**(4) 213–217, 1992.

Harris, R. J., "Gas explosions in buildings and heating plant," *Explosion and Fire News* Spon. (British Gas), London, 1983.

Harrison, A. J., and Eyre, J. A., "The effect of obstacle arrays on the combustion of large premixed gas/air clouds," *Combustion Jcl and Tech.,* **32**, 121–137, 1987.

Havens, J. A., "A Dispersion Model for Elevated Dense Gas Jet Chemical Releases," Volumes I and II, EPA-450/4-88-006, U. S. Environmental Protection Agency, 1988.

Havens, J.A., and Spicer, T.O., "LNG Vapor Dispersion Prediction with the DEGADIS Dense Cloud Dispersion Model," Gas and Research Institute Report, Contract 5086-252-1287 (September 1990) p. 27.

Hayashi, S., and Ohtani, T., "Limiting factor of flame propagation in low-volatility fuel clouds," *18th Int. Symp. on Combustion at U. of Waterloo, Canada,* Aug. 17–22, 1980, Combustion Institute, Pittsburgh, PA, 1981.

Hazardous Cargo Bulletin, "Asking for trouble," **11**(11), p. 86, Nov. 1990.

Heinhold, D. W., Paine, R. J., Walker, K. C., and Smith, D. G., "Evaluation of the AIRTOX dispersion algorithms using data from heavy gas field experiments," *Proc. 5th Conf. on Application of Air Poll. Meteorology,* AMS, Boston, 1986.

Henry, R. E., and Fauske, H. K., "The two-phase critical flow of one-component mixtures in nozzles, orifices, and short-tubes," *J. Heat Transfer, Trans. of the ASME,* 179–187, May 1971 (also Paper No. 70-WA/HT-5, same title).

Hesse, D. J., "A computational procedure for calculating the mass of flammable vapor in a neutrally buoyant cloud," *International Conference and Workshop on Modeling and Mitigating the Consequences of Accidental Releases of Hazardous Materials,* New Orleans, LA, May 20–24, 1991, pp. 511–528.

Hinze, J. D., "Fundamentals of the hydrodynamic mechanism of splitting in dispersion processes," *AIChE Journal,* **1** (4), 289–295, 1955.

Hosker, R.P. Jr., "Estimates of dry deposition and plume depletion over forests and grassland," in *Radioactive Contaminants in the Atmosphere*, Symposium Proceedings, International Atomic Energy Agency, Vienna, 1973, pp. 291–309.

Hosker, R. P. Jr., and Pendergrass, W. R., "Flow and Dispersion near Clusters of Buildings," NOAA Technical Memorandum ERL-ARL-153, June, 1987.

Hunt, J. C. R., Abell, C. J., Peterka, J. A., and Woo, H., "Kinetmatical studies of the flow around free or surface-mounted obstacles; applying topology to flow visualization," *J. Fluid Mechanics,* **86** (1), 179–200, 1977.

Irwin, J.S., and Cope, A.M., "A theoretical variation of the wind profile power law exponent as a function of surface roughness and stability," *Atmos Environment,* **13**, 1979, pp. 191–194.

Islitzer, N., and Slade, D. H., "Diffusion and Transport Experiments in Meteorology and Atomic Energy," USAEC Report TID-24190, pp. 117–118, Environmental Sciences Services Administration, Washington, DC, 1968.

Izumi, Y., "Kansas 1968 Field Program Data Report," AFCRL-72-0041, Air Force Cambridge Research Laboratory (LYB), L.G. Hansom Field, Bedford, MA 1971.

Jann, P. R., "Evaluation of temporary safe havens," *J. Loss Prev. Process Ind.,* **2**, 33–39, 1989.

Jobson, D. A. "On the flow of compressible fluids through orifices," *Proc. Instn. Mech. Engrs.,* **169** (37), 767–776, 1955. See Chisholm (1983).

Jones, G. W., "Inflammation limits and their practical application in hazardous industrial operations," *Chemical Review,* **22**, 1938.

Jones, T. B., and Spracklen, C. T., "Ionospheric effects of the Flixborough disaster," *Nature,* 250, 719, Aug. 30, 1974.

Jones, T. B., "Ionospheric effects of the Flixborough explosion," Report to Flixborough Court of Inquiry (U. of Leicester) 1975.

Keagy, W.R., and Weller, A.E., "A Study of freely expanding inhomogeneous jets," *Heat Transfer and Fluid Mech Inst.,* **2**, 89–98, 1949.

Kingery, C. N., and Bulmash, G., "Air Blast Parameters from TNT Spherical Air Burst and Hemispherical Surface Burst," Tech. Rep. ARBRL-TR 02555, US Army Ballistic Res. Lab., Aberdeen Proving Ground, MD, 1964.

Kinsella, K. G., "A rapid assessment methodology for the prediction of vapor cloud explosion overpressure," *Proc. Int. Conf. on Safety, Health and Loss Prevention in the Oil, Chemical and Process Industries,* Singapore, February 1993.

Kletz, T. A., "Unconfined vapor cloud explosions," *AIChE Loss Prevention,* **11**, p. 50, 1977.

Kletz, T. A., *Critical Aspects of Safety and Loss Prevention,* Butterworths, Stoneham, MA. , 1990.

Koestel, A., Gido, R. G., and Lamkin, D. E., "Drop size estimates for a loss-of-coolant accident," NUREG/CR-1607, LA 8449, Washington, DC, August 1980.

Kohlen, R., Becker, C. D., and Ruppert, K. A., "Ein Beitrag zur Darstellung der Explosionsgrenzen Brennbarer Gase," Chem.-Ing.-Tech, MS 2034, 1992.

Kong, D., Eckhoff, R. K., and Alfert, F., "Auto-ignition of CH_4/air, C_3H_8/air, CH_4/C_3H_8/air and CH_4/CO_2/air using a 1 liter ignition bomb," *J. Haz. Materials,* **40**, 69–84, 1995.

Krawitz, A. A. et al., "Physical and Chemical Properties of Lubricants," Technical Document Report ML-PDR-64-27, Air Force Materials Laboratory, Research and Technology Division, Air Force Systems Command, Wright-Paterson Air Force Base, Ohio, March, 1964.

Kukkonen, J., and Vesala, T., "Aerosol dynamics in releases of hazardous materials," *International Conference and Workshop on Modeling and Mitigating the Consequences of Accidental Releases of Hazardous Materials,* New Orleans, LA, May 20–24, 1991, pp. 53–86.

Kunkel, B. A., "User's guide for the Air Force Toxic chemical dispersion model (AFTOX)," AFGL-TR-88-0009, Hanscom AFB, MA, 01731, 57 pp., 1988.

Lapina, R. P., "How to estimate compressibility factors and specific heat ratios for hydrocarbon gases," *Chemical Engineering,* pp. 90–91, February 8, 1982.

Laurendeau, N. M., "Thermal ignition of methane-air mixtures by hot surfaces: A critical examination," *Combustion & Flame,* **46**, 29–49, 1982.

Lechaudel, J. F., and Mouilleau, Y., "Assessment of an accidental vapor cloud explosion, A case study: Saint Herblain, October 7, 1991, France," *Loss Prev., and Safety Promotion in the Process Industries,* **1**, 377–388, 1995.

Lees, F. P., "*Loss Prevention in the Process Industries,* 1st Edition," Butterworth Heinemann, London, 1980, pp. 899–907.

Lees, F. P., "Loss Prevention in the Process Industries, 2nd Edition," Butterworth Heinemann, London, 1996.

Lenoir, E. M., and Davenport, J. A., "A survey of vapor cloud fires and explosions, second update," *Process Safety Progress,* **12** (1), 1993.

Leung, J. C., "A generalized correlation for one-component homogeneous equilibrium flashing choked flow," *AIChE J.*, **32** (10), 1743–1746, 1986.

Leung, J. C., and Grolmes, M. A., "The discharge of two-phase flashing flow in a horizontal duct," *AIChE J.*, **33** (3), 524–527, 1987.

Leung, J. C, "Similarity between flashing and nonflashing two-phase flows," *AIChE J.*, **36**, (5), 797–800, 1990.

Leung, J. C., "The omega method for discharge rate evaluation," *Int'l Symp. on Runaway Reactions and Pressure Relief Design,* AIChE, Boston, MA, pp. 367–393, Aug. 2–4, 1995.

Leung, J. C., and Epstein, M., "The discharge of two-phase flashing flow from an inclined duct," *J. Heat Trans., Trans. ASME,* **112**, 524–528, May, 1990.

Lewis, B., and von Elbe, G., "Combustion, Flames, and Explosion of Gases," Academic Press, New York, 1961.

Lewis, D. J., "Case studied of effects of liquid fuel fires and of vapor cloud fires and explosions," *The Risks of Fuel Transport,* Session 1, Paper 1, London, 1982.

Leyer, J. C., and Desbordes, D., "Unconfined deflagragrative explosion without turbulence: experiment and model," *J. Hazardous Materials*, **34**, (2) 123–150, 1993.

Lienhard, J.H., and Day, J.B., "The Breakup of Superheated Liquid Jets", J. *Basic Engineering, Trans AIME*, **515,** September 1970.

Lind, C. D., "What causes unconfined vapor cloud explosions –Historical perspective and predictive method based on incident records," *Prog. Energy Combustion Science,* **6**, 151, 1975; "What causes unconfined vapor cloud explosions?," *Loss Prevention, AIChE,* **9**, 101–105, 1975.

Liu, M.K., Durran, D., Mundkur, P., Yocke, M., and Ames, J., "The Chemistry, Dispersion, and Transport of Air Pollutants Emitted from Fossil Fuel Power Plants in California: Data Analysis and Emission Impact Model," Final Report Contract No. ARB4-258 Systems Application Inc., San Rafael, CA 1976.

Lo, A.K., and McBean, G.A., "On the relative errors in methods of flux calculation," *J. Applied Meteorology,* **17**, 1704–1711, 1978.

Magnus, W., Oberhettinger, F., and Soni, R. P., "*Formulas and Theorems for the Special Functions of Mathematical Physics,*" Springer-Verlag, New York, 1966, pp. 262–295.

Mahoney, D. G., *Large Property Damage Losses in the Hydrocarbon-Chemical Industries: A Thirty-Year Review,* 13th ed., Marsh & McLennan Protection Consultants, New York, 1990.

Marsh & McLennan Protection Consultants, "Large Property Damage Losses in the Hydrocarbon-Chemical Industries: A Thirty-Year Review," 14th ed., p. 12, Marsh & McLennan Protection Consultants, New York, 1992.

Marshall, J. A., Personal communication regarding explosion at Dow Chemical, Freeport, Texas, April 29, 1998.

Marshall, M. R., "The effect of ventilation on the accumulation and dispersal of hazardous gases," *Proceedings of the 4th Int. Symp. On Loss Prevention and Safety Promotion in the Process Industries,* Harrogate, Sept. 12–16, 1983; IChemE Symp. Ser. No. 81, Pergamon Press, Oxford, 1984.

Marshall, V. C., "The siting and construction of control buildings—a strategic approach," *Process Industry Hazards,* 187, 1976.

Marshall, V. C., *Major Chemical Hazards,* Ellis Horwood, Chichester, U.K., 1987.

McQuaid, J., "Climatological records from the Thorney Island mast," Heavy Gas Dispersion Trials Report, Health & Safety Executive, Research & Laboratory Services Division, June 1981.

McFarlane, K., Prothero, A., Puttock, J. S., Roberts, P. T., and Witlox, H. W. M., "Development and validation of atmospheric dispersion models for ideal gases and hydrogen fluoride, Part I: technical reference manual," Report number TNER 90.015, Shell Research Ltd., Thornton Research Center, Chester, UK, 1990.

Melhem, G. A., "Hazard reduction benefits from reduced storage temperature of pressurized liquids," in (V. Fthenakis, Ed.), *Prevention and Control of Accidental Releases of Hazardous Gases, Van Nostrand Reinhold, 1993, pp. 422–428.*

Mercx, W. P. M., Johnson, D. M., and Puttock, J. S., "Validation of scaling techniques for experimental vapor cloud explosion investigations," *Proc. AIChE Loss Prevention Symposium,* Atlanta, GA, April 1994.

Mercx, W. P. M., and van den Berg, A. C., "The explosion blast prediction model in the revised CPR 14E (*Yellow Book*)," *Process Safety Progress,* **16** (3), 152–159, 1997.

Mercx, W. P. M. (Project co-ordinator), "Extended modelling and experimental research into gas explosions. Final summary report for the project EMERGE," CEC contract EV5VCT930274, Undated (1998a?).

Mercx W.P.M. (Project co-ordinator), "Modelling and experimental research into gas explosions, overall final report of the MERGE project, CEC contract: STEP - CT 0111 (SSMA), Undated (1998b?).

Mercx, W. P. M., "Large scale experimental investigation into vapor cloud explosions, comparison with the small scale DISCOE trials," *7th International Symposium on Loss Prevention and Safety Promotion in the Process Industries,* Taormina, Italy, 4–8 May 1992.

Mercx, W. P. M., "International Course on Explosion Blast Modeling in the Revised *Yellow Book,*" Barcelona, Spain, May 4, 1998.

Mills, M., Techniques for modeling jet releases for emergency prevention and preparedness applications," 1988 Hazardous Materials Spills Conference, AIChE and the National Response Team, Chicago, IL, 1988.

Minstry of Social Affairs (The Netherlands), "Report on the explosion at DSM in Beek, November 7, 1975," The Hague, The Netherlands, 1976.

Mudan, K. S., "Evaluation of fire and flammabililty hazards," Encyclopedia of Environmental Control Technology," Vol. 1, Chapter 14, p. 416, Gulf Publishing Co., Houston, TX 1989.

Munday, G., "The sizes of possible flammable vapor cloud discharged from Section 25A" Report to Flixborough Court of Inquiry, Cremer and Warner Report 9, Imperial College, London, 1975.

Nettleton, M. A., *Gaseous Detonations, Their Nature, Effects, and Control,* Chapman and Hall, New York, 1987.

NFPA (National Fire Protection Association) 68, "Guide for Venting of Deflagrations," Quincy, MA, 1994.

NFPA (National Fire Protection Association), *Fire Protection Handbook*(A. E. Cope, Ed.), 17th edition, pp. 2-293 to 2-299, Quincy, MA, 1991.

NOAA (National Oceanic and Atmospheric Administration), "ALOHA—Areal Locations of Hazardous Atmospheres, Version 4.0 User's Manual," Hazardous Material Response Branch, NOAA, Seattle, WA 98115, 1988.

Nou, J. V., "The Ocean Breeze and Dry Gulch Diffusion Programs," AFCRL, Hanscom AFB, MA, 1963.

Ooms, G., Mahieu, A. P., and Zelit, F., "The plume path of vent gases heavier than air," *1st International Loss Prevention Symposium and Safety Promotion in the Process Industries* (C. H. Buschman Ed.), Elsevier, Amsterdam, 1974, pp. 211–219.

Opschoor, G., "Methoden voor het berekenen van de fysische effecten van het incidenteel vrijkomen van gevaarlijke stoffen, hoofdstuk 4: Spray release," Directoraat Generaal van de Arbeid, CPR 14, Voorburg, 1988.

Osakabe, M., and Isono, M., "Effect of valve lift and disk surface on two-phase critical flow at hot water relief valves" *Int. J. Heat & Mass Transfer,* **39** (8) 1617–1624, 1996.

Panofsky, H. A., and Dutton, J. A., *Atmospheric Turbulence,* John Wiley & Sons, New York, 1984, pp. 5–8.

Parker, R. J. (chairman), "The Flixborough disaster, report of the Court of Inquiry," HM Stationery Office, London, 1975.

Pasquill, F., *Atmospheric Diffusion,* 2nd ed., Halstead Press, New York, 1974.

Pasquill, F, "The estimation of dispersion of windborn material," *Meteorol. Mag.*, **90**, 33–49, 1961.

Pasquill, F., and F. B. Smith, "The physical and meteorological basis for the estimation of the dispersion of windborn material," in (H. M. Englund and W. T. Beery, Eds.), *Proceedings of the Second International Clean Air Congress,* Washington, DC, 1970, pp. 1067–1072, Academic Press, New York, 1971.

PERF. ...experiments on obstacle effects on dispersion, ca 1996–97 at Nevada test site.

Perry's Chemical Engineer's Handbook, D. W. Green, Ed., 7th ed., McGraw-Hill, New York, 1997.

Peterson, R. L., "Surface Roughness Effects on Heavier-Than-Air Gas Diffusion," API Publication No. 4459, American Petroleum Institute, Washington, DC, Oct. 1987.

Peterson, R. L., "Effect of homogeneous and heterogeneous surface roughness on heavier-than-air gas dispersion," *J. Wind Eng. and Indust. Aerodynamics,* **36**, 643–652, 1990.

Peterson, R. L., "Vapor barriers," in (V. M. Fthenakis, Ed.), *Prevention and Control of Accidental Releases of Hazardous Gases,* Van Nostrand Reinhold, New York, 1993, pp. 174–210.

Peterson, R. L. "A wind tunel evaluation for estimating surface roughness lengths at industrial facilities," *Atmos. Environment,* **31** (1) 45–57, 1997.

Peterson, R. L., and Ratcliff, M. A., "Effect of homogeneous and heterogeneous surface roughness on HTAG dispersion," CCP Project 87-0417, Cermak Peterka Peterson Inc., Fort Collins, CO, 80524, 1988.

Peterson, W. B., and Lavdas, L. G., "INPUFF 2.0—A Multiple Source Gaussian Puff Dispersion Algorithm," Users Guide, ASRL, USEPA, Research Triangle Park, NC, 1986.

Pettit, G., "Characterization of Two-Phase Releases," Doctoral dissertation, South Bank Polytechnic, London, 1990.

Pieterson, C. M., "Analysis of the LPG Incident in San Juan Ixhuatapec, Mexico City, 19 Nov. 1984," Report 85-0222, TNO, Apeldoor, The Netherlands, 1985.

Pieterson, C. M., "Analysis of the LPG disaster in Mexico City," *Loss Prev. and Safety Promotion,* **5**, 21-1, 1986.

Pieterson, C. M., "Analysis of the LPG disaster in Mexico City," *J. Hazardous Materials,* **20**, 85, 1988.

Press, W. H., Flannery, B. P., Teulolsky, S. A., and Vetterling, W. T., *Numerical Recipes,* Cambridge Univ. Press, Cambridge, U.K., 1986, pp. 183–190.

Prugh, R. W., "Evaluation of unconfined vapor cloud explosion hazards," *International Conference on Vapor Cloud Modeling*, Cambridge, MA, 712–755, 1987

Purdy, G., "Testing the effectiveness of emergency planning strategies for reducing the risk from the transportation of dangerous goods," *Dangerous Goods Emergency Response Conference, '89,* Halifax, Nova Scotia, Canada, May 16–18, 1989.

Puttock, J. S., "Fuel gas explosion guidelines—the congestion assessment method," *2nd European Conference on Major Hazards On and Off-shore,* Manchester, UK, Oct. 24, 1995.

Quest Consultants, Inc., "Focus Input and Reference Manuals," Norman, OK 73070, 1990.

Radian Corporation, "CHARM Emergency Response System Documentation," Austin, TX 78720, 1991.

Rae, D., Singh, B., and Danson, R., "The size and temperature of a hot square in a cold surface necessary for the ignition of methane," Research Report 224, Safety in Mines Res. Estab., Buxton, UK, 1964.

Raj, P. K., and Emmons, H. W., "On the burning of a large flammable vapor cloud," Western and Central States Section Meeting of Combustion Institute, San Antonio, TX, 1975.

Raj, P. K., Mudan, K. S., and Moussa, N. A., "Experiments Involving Pool and Vapor Fires from Spills of Liquefied Natural Gas on Water," US Coast Guard Report CG-D-55-79, 1979.

Raj, P. K., and Morris, J. A., "Source Characteristics and Heavy Gas Dispersion Models for Reactive Chemicals," AFGL-TR-88-0003 (1), Hanscom AFB, MA 01730, 1987.

Rajaratnam, N. *Turbulent Jets,* Elsevier, New York, 1976, pp. 184–202.

Rew, P. J., Deaves, D. M.and Maddison, T., "Current status and advances in flash fire modeling," *International Conference, and Workshop on Modeling and Mitigating the Consequences of Accidental Releases of Hazardous Materials,*" New Orleans, LA, Sept. 26–29, 1995, AIChE, New York, 1995, pp. 737–761.

Ricou, F. P., and Spalding, D. B., "Measurements of entrainment by axisymmetrical turbulent jets," *J. Fluid Mech.,* **7,** 21–32, 1962.

Roberts, P. T., Puttock, J. S., and Blewitt, D. N., "Gravity spreading and surface roughness effects in the dispersion of dense gas plumes," *AIChE Health and Safety Symposium,* Orlando, Florida, March 18–22, 1990

Rota, R., Morbidelli, M., Levresse, B., and Ditali, S., "Simulation of short duration gas releases," *Proceedings of the 8th International Symposium on Loss Prevention and Safety Promotion in the Process Industries,* Vol. II, pp. 303–314, Elsevier, Amsterdam, 1995.

Sachs, R. G., "The Dependence of Blast on Ambient Pressure and Temperature," BRL Report No. 466, Aberdeen Proving Ground, MD, 1944.

Sadee, C., Samuels, D. E., and O'Brien, T. P., "The characteristics of the explosion of cyclohexane at the New Yorkpro (UK) Flixborough plant on June 1st 1974," J. *Occup. Accid.,* **1**, 203, 1976–77.

SAFER, Inc., "User's Manual for TRACE II," Westlake Village, CA 91362, 1989.

Sedefian, L., and Bennett, E., "A comparison of turbulence classification schemes," *Atmospheric Environment,* **14** (7) 741–750, 1980.

Schatzmann, M., Marotzke, K., and Donat, J., "Research on continuous and instantaneous heavy gas clouds—contribution of sub-project EV4T-00210D to the final report of the joint CEC project," Univ. of Hamburg, Meteorological Institute Report, 1990.

Schmidli, J., "The Initial Phase of Sudden Releases of Superheated Liquid," Doctoral dissertation, Swiss Federal Institute of Technology Zurich, 1993.

Schumann, S., and Wirkner-Bott, I., "Vapor Cloud Explosion After Two-Phase Jet Release," Battelle Europe report, 36-1 to 36-29, 1991?.

Shair, F. H., and Heitner, K. L., "Theoretical model for relating indoor pollutant concentrations to those outside," *Environmental Sci. and Technology,* **8** (5) 444–451, May 1974.

Shaw, S. "Bangkok LPG truck disaster 24th September 1990," *Loss Prevention Bulletin,* no. 101, 18–22, Oct. 1991.

Shepherd, J. E., Melhem, G. A., and Athens, P., "Unconfined vapor cloud explosions: a new perspective," *International Conference and Workshop on Modeling and Mitigating the Consequences of Accidental Releases of Hazardous Materials,*" New Orleans, LA, May 20–24, 1991, AIChE, New York, 1991, pp. 613–635.

Simmons, J. A., "Risk Assessment or Storage and Transport of liquefied Natural Gas and LP Gas," Final report for Contract 68.01.2695 for US Environmental Protection Agency, 1974.

Simpson, L. L. "Navigating the two-phase maze," International Symposium on Runaway Reactions and Pressure Relief Design, (G. A. Melhem and H. G. Fisher, Eds.), Boston, MA, Aug. 2–4, 1995, AIChE, New York, 1995, pp. 394–417.

Skandia International Insurance Co., "BLEVE! The tragedy of San Juanico," *First newsletter Journal,* 1985.

SMEDIS Project, "Document Control and Classification for SMEDIS (Scientific Method for Evaluation of Dispersion Models," CERC Ltd., 3 King's Parade, Cambridge, UK CB2 1SJ, January 7, 1997.

Snee, T. J., "Auto-ignition, slow combustion and the variation of minimum ignition temperature with vessel size," *IChemE Loss Prevention Bulletin* **81**, 25–38, 1988.

Spencer, H., and Rew, P. J., "Ignition probability of flammable gases," HSE Books, Her Majesty's Stationery Office, Norwich, UK, 1997, p. 28.

Spicer, T. O., and Havens, J. A., "User's Guide for the DEGADIS 2.1 Dense Gas Dispersion model," Environmental Protection Agency Report EPA-450/4-89-019, 1989.

Spicer, T. O., and Havens, J. A., "Application of dispersion models to flammable cloud analyses," *J. Hazardous Materials,* **49**, 115–124, 1996.

Sommerfeld, J. T., "Tank draining revisited," *Chem. Eng.*, 171–172, May 1990.

Stoc, M., Schumann, St., and Wirkner-Bott, I., "Vapor cloud explosion after two-phase jet release," Report by Battelle Europe, Frankfort, GermaNew York, Chapter 36, 1991?.

Stull, D. R., and R. Prophet, Eds. *JANAF Thermochemical Tables,* 2nd ed., MBS-37, June, 1971.

Tamanini, F., "An improved correlation of experimental data on the effects of ducts in vented dust explosions," *Loss Prevention and Safety Promotion in the Process Industries,* **1**, 243–253, 1995.

Tamanini, F., and Valiulis, J. V., "Improved guidelines for sizing of vents in dust explosions," *J. Loss Prev. Process Ind.*, **9**(1) 105–118, 1996a.

Tamanini, F., "Vent sizing in partial-volume deflagrations and its application to the case of spray dryers," *J. Loss Prev. Process Ind.*, **9**(5), 339–350, 1996b.

Tamanini, F, "Modelling of panel inertia effects in vented dust explosions," *Process Safety Progress,* **15**(4) 247–257, 1996c.

Tamanini, F, "The use of models in the development of explosion protection guidelines," *9th International Symposium on Loss Prevention and Safety Promotion in the Process Industries,* Barcelona, Spain, 4–7 May 1998.

The Netherlands, Labor Inspectorate, "Guidelines on Designation of Installations for Which a Safety Report Is Mandatory," Publication P172-1E , 1990?.

Tickle, G. A., "A model of the motion and dilution of a heavy gas cloud released on a uniform slope in calm conditions," *J. Hazardous Materials,* **49**, 29–47, 1996.

TNO, Committee for the Prevention of Disasters, "Methods for the calculation of physical effects," sponsored by the Director-General for Social Affairs and Employment, The Hague, The Netherlands, *Yellow Book,* 1997.

Touma, J.S., "Dependence of the wind profile power law in stability at various locations," *J Air Pollut. Cont. Assoc.*, **27**, 863–866, 1977.

U.S. Dept. of Labor, "Phillips 66 CompaNew York Houston Chemical Complex Explosion and Fire, A Report to the President," April 26, 1990.

VDI 3673, *Pressure Release of Dust Explosions,* Verein Deutscher Ingenieure, Nov. 1995.

van den Berg, A. C., "The multi-energy method—a framework for vapor cloud exposion blast prediction," *J. Hazardous Materials,* **12**, 1–10, 1985.

van den Berg, A. C., and Eggen, J. B. M. M., "GAME: Guidance for the Application of the Multi-Energy Method," *2nd Int. Specialists Meeting on Fuel–Air Explosions*, Bergen, Norway, June 1996.

Van Buijtenen, C. J. P., "Calculation of the amount of gas in the explosive region of a vapor cloud released in the atmosphere," *J. Hazardous Materials,* **3**, 201–220, 1980.

Vanderstraeten, B., Tuerlinckz, D., Berghmans, J., Vliegen, S., Van't Oost, E., and Smit, B., "Experimental study of the pressure and temperature dependence on the upper flammabililty limit of methane/air mixtures," *J. Hazardous Materials,* **56**, 237–246, 1997.

Van Wingerden, K., Salvesen H-C., and Perbal, R., "Simulation of an accidental vapor cloud explosion," *Process Safety Progress,* **14**(3), 173–181, 1995.

Vesala, T., Kukkonen J., and Kulmala M., "A Model for Evaporation of Freely Falling Droplets," Publications on Air Quality, Finnish Meteorological Institute, Helsinki, Finland, 1989.

Venart, J. E. S. "Flixborough: A final resolution?" *Interflame-99,* Edinburgh, U. K., July 1, 1999.

Webber, D. M., Jones, S. J., Tickle, G. A., and Wren, T., "A Model of a Dispersing Dense Gas Cloud and the Computer Implementation D*R*I*F*T. I: Near Instantaneous Release." SRD Report SRD/HSE R586 April 1992, and " II: Steady Continuous Releases." SRD Report SRD/HSE R587 April 1992

Webber, D. M., Jones, S. J., and Martin, D., *J. Hazardous Materials,* **33**, 101, 1993.

White, A. G., "Limits for the propagation of flame in vapor–air mixtures; I. Mixtures of air and one vapor at ordinary temperatures and presssure," *J. Chem. Soc.,* **121**, 1922.

Wiekema, B. J., "Vapor cloud incidents—an analysis based on accidents," *J. Hazardous Materials,* **8**, 295, 313, 1984.

Williams, F. A.,"*Combustion Theory,* 2nd ed., Benjamin/Cummings, Menlo Park, CA, 1985.

Wilson, D. J., "The Release and Dispersion of Gas from Pipeline Ruptures," report to Alberta Environment, Contract No. 790686, Edmonton, Alberta, Canada, 1979.

Wilson, D. J., "Expansion and Plume Rise of Gas Jets from High Pressure Pipeline Ruptures," report to Alberta Environment, Edmonton, Alberta, Canada, Contract No. 810786, April, 1981.

Wilson, D. J., "Variation of indoor shelter effectiveness caused by air leakage variability of houses in Canada and the USA," *U.S. EPA/FEMA Conference on the Effective Use of in-Place Sheltering as a Potential Option to Evacuation during Chemical Release Emergencies,* Emmitsburg, MD, Nov. 30 and Dec. 1, 1988.

Wilson, D. J., "Effectiveness of indoor sheltering during long duration toxic gas releases," *ER'91—Emergency Response Conference on Technological Response to Dangerous Substances Accidents,* Calgary, Alberta, Canada, May 28–30, 1991.

Wilson, D. J., *Concentration Fluctuations and Averaging Time in Vapor Clouds,* CCPS, AIChE, New York, 1995, pp. 29–30.

Wilson, D. J., "Do leaky buildings provide effective shelter?," *Proceedings of the 10th Annual Conference on Major Industrial Accidents,* Council of Canada, Edmonton, Alberta, Canada, Nov. 4–7, 1996.

Witlox, H. W. M., "User's Guide for the HEGADAS Heavy Gas Dispersion Program," Shell Research Ltd., 1988.

Witlox, H. W. M., McFarlane, K., Rees, F. J., and Puttock, J. S., "Development and Validation of Atmospheric Dispersion Models for Ideal Gases and Hydrogen Fluoride, Part II, HGSYSTEM Program User's Manual," Shell Int. Research, Maatschappij, The Netherlands, 1990.

Woo, H. G. C., Peterka, J. A., and Cermak, J. E., "Wind-Tunnel Measurements in the Wakes of Structures," NASA Contractor Report NASA CR-2806, 226 pp., 1977.

Woodward, J. L. "Discharge rates through holes in process vessels and piping," in (V. M. Fthenakis, Ed.), *Prevention and Control of Accidental Releases of Hazardous Gases,* Van Nostrand Reinhold, New York, 1993, pp. 94–159.

Woodward, J. L., "Aerosol drop size correlation and corrected rainout data using models of drop evaporation, pool absorption, and pool evaporation," International Conference and Workshop on Modeling and Mitigating the Consequences of Accidental Releases of Hazardous Materials, New Orleans, LA, Sept. 26–29, 1995, AIChE, New York, 1995a.

Woodward, J. L. "An amended method for calculating omega for a homogeneous equilibrium model of predicting discharge rates," J. Loss Prevention in Process Industries, " **8** (5), 253–259, 1995b.

Woodward, J. L., and Cook, J., "Tuning a complex suite of dispersion models," *International Conference and Workshop on Modeling and Mitigating the Consequences of Accidental Releases of Hazardous Materials,* New Orleans, LA, Sept. 26–29, 1995, AIChE, New York, 1995.

Woodward, J. L., Cook J., and Papadourakis, A., "Modeling and validation of a dispersing aerosol jet," *J. Hazardous Materials*, **44**, 185–207, 1995.

Woodward, J. L., and Mudan, K. S., "Liquid and gas discharge rates through holes in process vessels," *J. Loss Prev. Process Ind.,* **4**(3), 161–165, 1991.

Woodward, J. L., and Papadourakis, A., "Modeling of droplet size, evaporation, and rainout in a dispersing aerosol jet", *International Conference and Exhibition on Safety, Health, and Loss Prevention in the Oil, Chemical and Process Industries,* Singapore, Feb. 15–19, 1993.

Woodward, J. L., and Papadourakis, A., "Reassessment and reevaluation of rainout and drop size correlation for an aerosol jet," *J. Hazardous Materials*, **44**, 209–230, 1995.

World Map of Natural Hazards, Munich, Germany, 1995.

Wu, J. M. C., and Schroy, J. M., "Emissions from spills," *Proceedings of APCA/WPCF Joint Specialty Conference: Control of Specific Toxic Pollutants,* Gainseville, Florida, APCA, Feb. 1979.

Zabetakis, M. G., "Flammability characteristics of gases and vapors," *Bureau of Mines Bulletin* 627, 1965a.

Zabetakis, M. G., "Fire and Explosion Hazards at Temperature and Pressure Extremes," *AIChE –I. Chem. E. Symposium Series* No. 2, pp. 99–104, 1965b.

Zabetakis, M. G., Lambiris, and Scott (1959).

Zeeuwen, J. P., and B. J. Wiekema, "The measurement of relative reactivities of combustible gases," *Conference on Mechanisms of Explosion in Dispersed Energetic Materials,*" 1978.

INDEX

Publications Available from the
CENTER FOR CHEMICAL PROCESS SAFETY
of the
AMERICAN INSTITUTE OF CHEMICAL ENGINEERS
3 Park Avenue, New York, NY 10016-5901

CCPS Guidelines Series

Guidelines for Pressure Relief and Effluent Handling Systems
Guidelines for Design Solutions for Process Equipment Failures
Guidelines for Safe Warehousing of Chemicals
Guidelines for Postrelease Mitigation in the Chemical Process Industry
Guidelines for Integrating Process Safety Management, Environment, Safety, Health, and Quality
Guidelines for Use of Vapor Cloud Dispersion Models, Second Edition
Guidelines for Evaluating Process Plant Buildings for External Explosions and Fires
Guidelines for Writing Effective Operations and Maintenance Procedures Guidelines for Chemical Transportation Risk Analysis
Guidelines for Safe Storage and Handling of Reactive Materials
Guidelines for Technical Planning for On-Site Emergencies
Guidelines for Process Safety Documentation
Guidelines for Safe Process Operations and Maintenance
Guidelines for Process Safety Fundamentals in General Plant Operations
Guidelines for Chemical Reactivity Evaluation and Application to Process Design
Tools for Making Acute Risk Decisions with Chemical Process Safety Applications
Guidelines for Preventing Human Error in Process Safety
Guidelines for Evaluating the Characteristics of Vapor Cloud Explosions, Flash Fires, and BLEVEs
Guidelines for Implementing Process Safety Management Systems Guidelines for Safe Automation of Chemical Processes
Guidelines for Engineering Design for Process Safety
Guidelines for Auditing Process Safety Management Systems
Guidelines for Investigating Chemical Process Incidents
Guidelines for Hazard Evaluation Procedures, Second Edition with Worked Examples
Plant Guidelines for Technical Management of Chemical Process Safety, Revised Edition
Guidelines for Technical Management of Chemical Process Safety

Guidelines for Chemical Process Quantitative Risk Analysis
Guidelines for Process Equipment Reliability Data with Data Tables
Guidelines for Safe Storage and Handling of High Toxic Hazard Materials
Guidelines for Vapor Release Mitigation

CCPS Concepts Series

Avoiding Static Ignition Hazards in Chemical Operations
Estimating the Flammable Mass of a Vapor Cloud
RELEASE: A Model with Data to Predict Aerosol Rainout in Accidental Releases
Local Emergency Planning Committee Guidebook: Understanding the EPA Risk Management Program Rule
Inherently Safer Cemical Processes. A Life-Cycle Approach
Contractor and Client Relations to Assure Process Safety
Understanding Atmospheric Dispersion of Accidental Releases
Expert Systems in Process Safety
Concentration Fluctuations and Averaging Time in Vapor Clouds

Proceedings and Other Publications

Proceedings of the International Conference and Workshop on Reliability and Risk Management, 1998
Proceedings of the International Conference and Workshop on Risk Analysis in Process Safety, 1997
Proceedings of the International Conference and Workshop on Process Safety Management and Inherently Safer Processes, 1996
Proceedings of the International Conference and Workshop on Modeling and Mitigating the Consequences of Accidental Releases of Hazardous Materials, 1995
Proceedings of the International Symposium and Workshop on Safe Chemical Process Automation, 1994
Proceedings of the International Process Safety Management Conference and Workshop, 1993
Proceedings of the International Conference on Hazard Identification and Risk Analysis, Human Factors, and Human Reliability in Process Safety, 1992
Proceedings of the International Conference and Workshop on Modeling and Mitigating the Consequences of Accidental Releases of Hazardous Materials, 1991
Safety, Health and Loss Prevention in Chemical Processes: Problems for Undergraduate Engineering Curricula

Printed and bound in the UK by
CPI Antony Rowe, Eastbourne